PROJECTS
FOR THE
ROUTER

Woodworking by
CASEY CHAFFIN

Text and
Photography by
NICK ENGLER

Popular
Science

 Sterling Publishing Co., Inc.   New York

Library of Congress Cataloging-in-Publication Data

Chaffin , Casey.
    Projects for the router / woodworking by Casey Chaffin ; text and
photography by Nick Engler.
       p. cm.
    Includes index.
    ISBN 0-8069-6680-7 (pbk.)
    1. Routers (Tools) 2. Woodwork. I. Engler, Nick. II. Title.
TT203.5.C47 1988
684'.083—dc19   87-26735
                                             CIP

3 5 7 9 10 8 6 4

Published in 1988 by Sterling Publishing Co., Inc.
387 Park Avenue South, New York, N.Y. 10016
Originally published by Grolier Book Clubs, Inc.
Copyright © 1987 by Casey Chaffin and Nick Engler
Distributed in Canada by Sterling Publishing
% Canadian Manda Group, P.O. Box 920, Station U
Toronto, Ontario, Canada M8Z5P9
Distributed in Great Britain and Europe by Casell PLC
Artillery House, Artillery Row, London SWIP IRT, England
Distributed in Australia by Capricorn Ltd.
P.O. Box 665, Lane Cove, NSW 2066
Manufactured in the United States of America
All rights reserved
Sterling ISBN 0-8069-6680-7 (pbk.)

Contents

The Many Uses of the Router

When attached to a worktable, the router becomes a shaper.

Most woodworkers think of the router as a fairly recent innovation. The last twenty years has seen its growth from an obscure cabinetmaker's tool to a common fixture in the home workshop. It's true that woodworkers have only recently begun to discover the versatility of the router, but the router may be the oldest portable power tool, predating the electric drill by several years.

The first crude router was made by R.L. Carter, a pattern maker, during the First World War. Carter was making a wooden pattern for a boiler casting, and had to round over the edges of sixteen separate 'cores' in the casting. The radii of all the edges had to be precisely the same.

To do the job with a spokeshave would have taken days of tedious carving. Instead, Carter stripped down an old barber's electric clipper so that he had just the motor and the worm gear. He ground the worm gear to make a radius cutter, and rigged some guides. With this setup, he finished the job in two hours.

Workmen from Carter's shop talked to local cabinetmakers about Carter's invention. Realizing the potential of this new tool, the cabinetmakers offered to buy it if Carter would produce it. After the armistice, Carter went into the portable power tool business. Ten years later, the R.L. Carter Company had produced more than 100,000 'routers'.

The first use of the router was for **shaping**, and this remains a major use today. Woodworking tool catalogues are full of dozens of shaper bits you can use to cut beads, coves, chamfers, and ogees in the edge of a board, or in framing or molding stock. Mount the router to the underside of a worktable, and you have a very capable shaper.

Woodworkers quickly found that if you can rig some way to guide the router—or guide the workpiece over the router bit—you can make accurate, repetitive cuts. This is called **template routing.** There are two common ways to guide the router around a template. If you're using a hand-held router, mount a guide bushing to the base. If your router is mounted to a table, use a pin, mounted to an arm, just above the bit. This second arrangement is also known as 'pin routing'.

Perhaps the most useful chore that the router can perform is cutting **joinery.** With just three types of bits (straight, rabbeting, and dovetail), you can cut an amazing number of joints—dadoes, grooves, rabbets, mortises, tenons, and several types of dovetails. Furthermore, the router often makes a smoother cut than other joinery tools. This, in turn, provides a better gluing surface.

The projects in this book utilize all the capabilities of the router—shaping, template routing, and joinery. Some may use only one of these capabilities, while others use all three. The router isn't the only tool you'll need to build these projects, of course. But after you build one or two, you'll find that the router is much more versatile than you may have suspected.

Nick Engler
May, 1987

There are a large variety of bits available to help you make your own frames and molding.

A guide bushing, attached to the base, guides the router along a template.

With a dovetail bit, you can cut both half-blind dovetails and french dovetails.

Super Router Table

The router is a wonderfully versatile tool—with the proper accessories. By itself, it's nothing more than a motor, a chuck, and a base. But with the proper accessories, it can be both a portable *and* a stationary power tool, each with dozens of uses.

The most important accessory for a router is a router table. It holds the router stationary, and leaves you free to manipulate the stock with both hands. With this table, we show four optional attachments that further expand the capabilities of the router. The *fence* makes the router into a shaper, and serves as a guide for other operations. The *pin arm* serves as a starter pin for freehand shaping, and, with the use of templates, makes it possible to do precise duplicate cuts. The *circle-cutting pin* lets you cut, shape, or make grooves and rabbets in circular shapes. And finally, with the *spindle-holding jig* you can turn spindles, cuts flutes in spindles, or make joints in cylindrical stock.

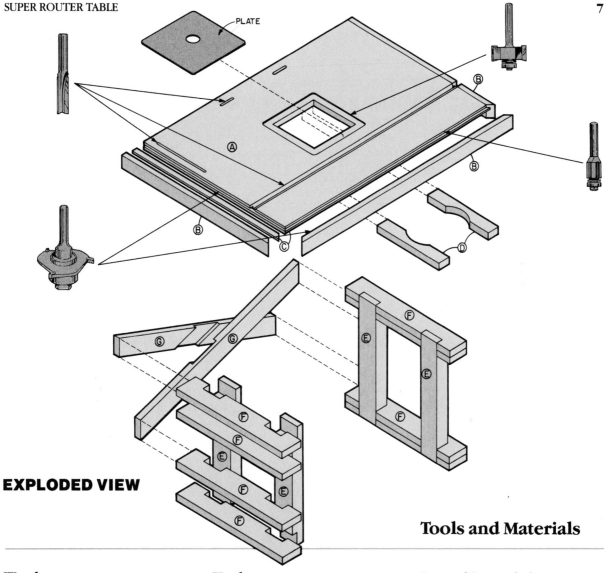

PLATE

Ⓐ Ⓑ Ⓒ Ⓓ Ⓔ Ⓕ Ⓖ

EXPLODED VIEW

Tools and Materials

Wooden parts:

A. Top   ¾" x 20" x 28¾"
B. Trim (total)   ¾" x 1½" x 72"
C. Splines (total)   ¼" x ¾" x 72"
D. Ribs (2)   ¾" x 1½" x 9"
E. Legs (4)   1½" x 1½" x 14"
F. Spreaders (8)   ¾" x 2⅜" x 15"
G. Braces (2)   ¾" x 2" x 24½"

Hardware:

- ³/₁₆" x 7½" x 7½" Aluminum mounting plate
- #8 x 1¼" Flathead wood screws (18-24)

Router bits needed:

- Slotting cutter bit
- Rabbeting bit
- ¼" and ½" Straight bits

Making the Router Table

---

1. Attach the router to the mounting plate.

Buy a mounting plate for your router. There are several commercial plates available from mail order companies, or you can make your own from $3/16''$ thick aluminum stock. Remove the base from the router and carefully center it on the plate. Using the base as a template, mark the mounting holes on the plate. (See Figure 1.) Drill and countersink these holes.

---

2. Trim the worktable.

Buy a laminated 'sink cutout' to make your worktable. With a slotting cutter, cut grooves for splines in the edge of the worktable and in the trim stock. (See Figure 2.) Cut and miter the trim, then attach the trim to the edge of the table with splines and glue.

---

3. Mount the plate to the table.

With a saber saw, cut a 6″ x 6″ opening in the middle of the table. With a rabbeting bit, cut a $3/16''$ deep, $3/4''$ wide rabbet all the way around the top edge of the opening. Clamp straight scraps of wood to the table to guide the router. (See Figure 3.) Glue the ribs to the underside of the table, on either side of the opening. Place the plate in the rabbet so that it's flush with the top surface, and screw it to the worktable and the ribs.

Figure 1. *Center the router base on the mounting plate, and mark where to drill the mounting holes.*

Figure 2. *Using a slotting cutter, make grooves in the worktable and the trim for the splines.*

Drill ¼″ holes to mark the beginnings and the ends of the slots along the back and the sides of the table, as shown in the working drawings. Using a straightedge to guide the router, cut the slots with a ¼″ straight bit. (See Figure 4.) If you wish, you can also make a groove along the front edge of your table, to fit the miter gauge from your table saw. Use a ½″ straight bit and cut the groove in several passes.

4. **Cut the slots and the groove in the table.**

Cut the legs and the spreaders, and notch them on a bandsaw. Drill and counterbore the top spreaders for the screws that you will use to attach them to the worktable. Assemble the legs and spreaders with glue and screws to make two trestles.

5. **Make the trestles.**

Make the parts for the crossbrace, then cut the lap joint on your table saw. Attach the crossbrace to the trestles with glue and screws. Then attach the trestles to the worktable with screws. *Don't* glue the work-table in place, in case you need to replace it someday.

6. **Attach the trestles to the table.**

Figure 3.   With a rabbeting bit, cut the rabbet for the mounting plate. Clamp straight pieces of wood to the worktable to guide the router.

Figure 4.   To cut the slots in the worktable, first drill holes to make the ends of the slots. Cut the slots with a ¼″ straight bit, using a straightedge as a guide.

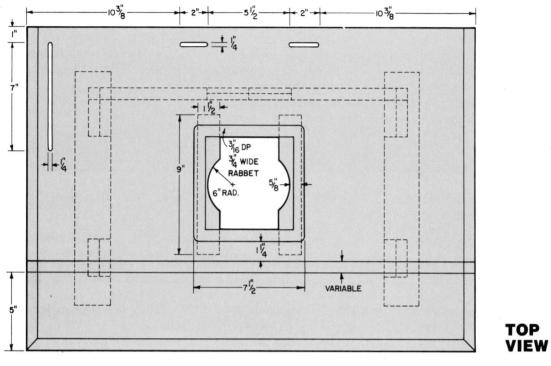

TOP VIEW

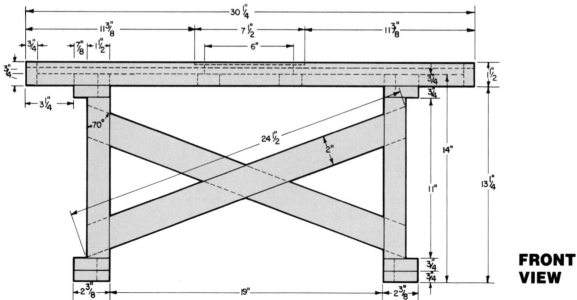

FRONT VIEW

Wire the underside of the worktable with a grounded plug and cord, an on/off switch, and an outlet, as shown in Figure 5. This arrangement will help you turn the router on and off easily, when using the Router Table.

7. Wire the completed Router Table.

Figure 5. Wire the underside of the worktable with an outlet, a switch, and a grounded plug and cord. This will allow you to easily turn the router on and off, when it's attached to the Router Table.

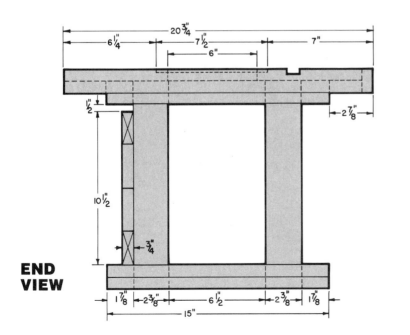

END VIEW

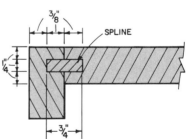

SPLINE

EDGE TRIM DETAIL

Making the Attachments

8. **Make a fence attachment.**

The *fence* is the most important attachment for your Router Table. It attaches to the two grooves along the sides of the worktable. You can adjust its position to help guide stock when you're edge-shaping or cutting joinery. Make the cutout in the middle of the fence with a ½″ straight bit. (See Figure 6.)

Figure 6. *Make the cutout in the middle of the fence with a ½″ straight bit. Cut the opening in several passes, removing ⅛″-¼″ more stock with each pass.*

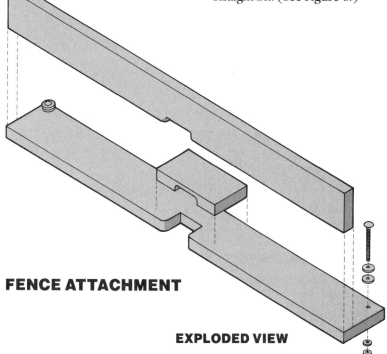

FENCE ATTACHMENT

EXPLODED VIEW

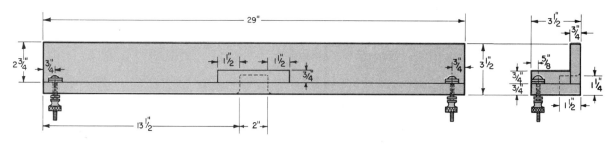

FRONT VIEW   **SIDE VIEW**

The *pin arm* provides a starting pin for freehand shaping, or a pilot for cutting rabbets or shaping irregular edges. It can also be used for 'template routing'. With the pin positioned just above the router bit, the pin will follow a template and guide the stock over the bit. The pin arm attaches to the slots along the back edge of the table, and can be raised or lowered by means of slots in the support bracket. To make these slots in the bracket, first drill starting and stopping holes, as you did when you made the slots in the table. Cut the slots with a ¼″ straight bit, using the fence attachment as a guide. (See Figure 7.)

9. **Make a pin arm attachment.**

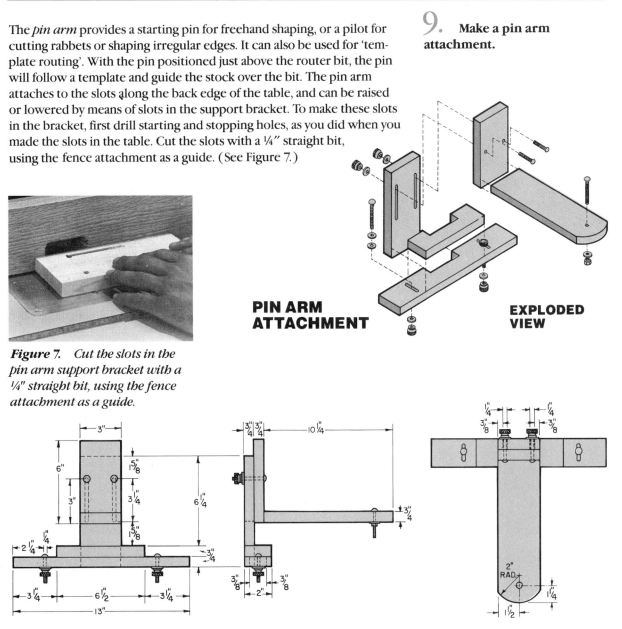

PIN ARM ATTACHMENT

EXPLODED VIEW

Figure 7. Cut the slots in the pin arm support bracket with a ¼" straight bit, using the fence attachment as a guide.

FRONT VIEW

SIDE VIEW

TOP VIEW

10. Make a circle-cutting attachment.

The *circle cutter* will cut both circular workpieces and circular openings in workpieces. The pivot pin can be repositioned either by moving it to different holes, or by sliding the attachment back and forth in the miter gauge slot. To lock the attachment in place, simply clamp it to the table.

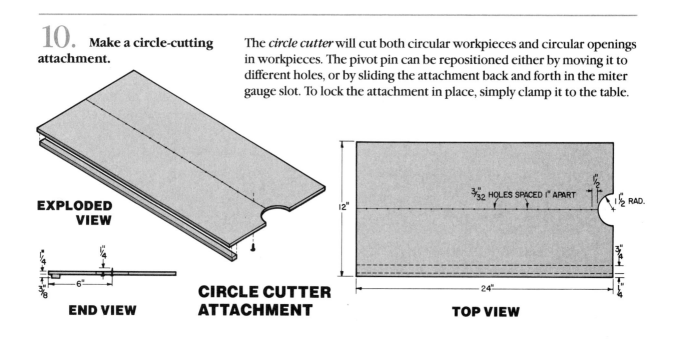

EXPLODED VIEW

END VIEW

CIRCLE CUTTER ATTACHMENT

TOP VIEW

11. Make a spindle-holding attachment.

The *spindle holder* will hold wood so that you can use your router as a lathe. It can also be used, along with the fence attachment, to cut flutes and joinery in spindles. The stock pivots between two centers. On one center, there is an indexing wheel to prevent the wood from turning while you rout it, and to measure just how far you turn the wood. This wheel is removable; you can make a new one for each project. Make the slots and the other joinery in this attachment using straight bits. (See Figure 8.) Be particularly careful when assembling this attachment; *accuracy is critical.*

Figure 8. *Cut the rabbets in the spindle holder spreaders with a ½" straight bit. Make the joint in several passes, using a miter gauge to help guide the work.*

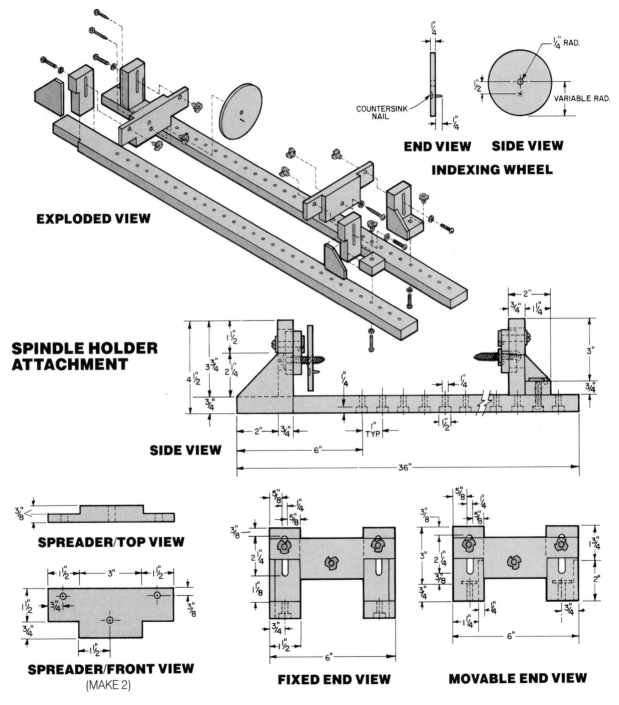

EXPLODED VIEW

END VIEW **SIDE VIEW**

INDEXING WHEEL

COUNTERSINK NAIL

¼" RAD.

VARIABLE RAD.

SPINDLE HOLDER ATTACHMENT

SIDE VIEW

SPREADER/TOP VIEW

SPREADER/FRONT VIEW
(MAKE 2)

FIXED END VIEW

MOVABLE END VIEW

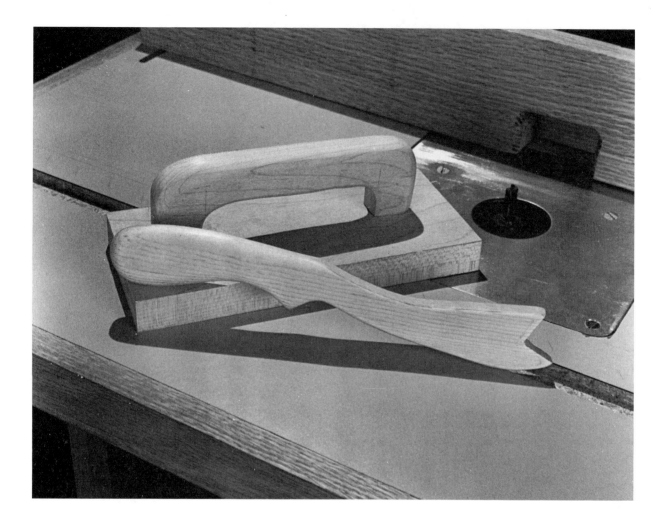

Safety Tools

To use your Router Table safely, you need to keep your hands and fingers away from the whirling bits. From time to time, you may be tempted to shape or cut a small wooden part that will bring your fingers dangerously close to the bit. Don't do it! The bit may catch and pull the part from your hands, and if you're putting even a slight forward pressure on the piece, you can push your fingers into the bit before you know what's happening.

Instead of using your hands to feed the smaller pieces of wood into the bits, use the safety tools you see here—push sticks and push shoes. By using them, you can shape those small parts and still keep your hands a reasonable distance from the bit.

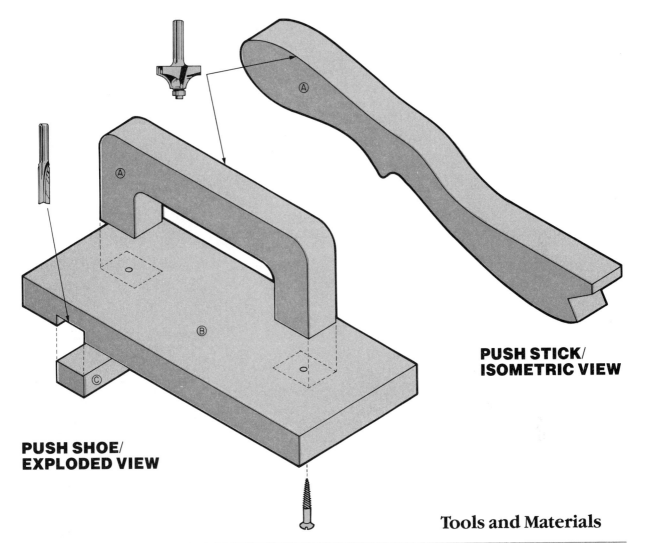

**PUSH STICK/
ISOMETRIC VIEW**

**PUSH SHOE/
EXPLODED VIEW**

Tools and Materials

Wooden parts:

Push Stick
A. Push stick   ¾″ x 2½″ x 10″

Push Shoe
A. Handle   ¾″ x 2⅛″ x 6″
B. Base   ¾″ x 3½″ x 8″
C. Heel   ½″ x ¾″ x 3½″

Hardware:

- #10 x 1½″ Flathead wood screws (2)

Router bits needed:

- ¼″ Piloted quarter-round bit
- ¾″ Straight bit

Making the Push Stick

1. **Select the stock.**

Select a scrap of clear wood to make the push stick. The wood must be free of all defects—no checks, splits, or knots. These defects could cause the stick to break while you're using it. In addition, the wood should not splinter easily. For this reason, don't use plywood.

2. **Cut out the push stick pattern.**

Mark the push stick pattern on the stock. Then cut it out with a bandsaw or jigsaw.

3. **Round over the edges of the handle.**

Mount a piloted ¼″ quarter-round bit in your router. Round over all the edges of the handle, where you'll hold the push stick. (See Figure 1.) You needn't round the shank of the tip.

TIP While you're at it, make half a dozen or so push sticks. Keep them near each of the tools in your shop where they might come in handy—router table, bandsaw, table saw, jointer, etc.

Figure 1.   Using a piloted quarter-round bit, round over the handle of the push stick.

Making the Push Shoe

Cut the stock for the handle, base, and heel. The heel on this tool is optional; some people prefer to use push shoes without the heel. (A push shoe without a heel is commonly called a 'push block'.) Depending on the type of woodworking you do, it may be a good idea to make a push shoe *and* a push block—with and without the heel. There will be times when the heel is indispensable, and other times you'll find it just gets in your way.

1. **Cut the stock to size.**

Bevel the lower edge of the handle stock at 10°, so that you can later mount the handle at a slight angle. Then lay out the shape of the handle on the stock, and cut it out on a bandsaw.

2. **Make the handle.**

I SQUARE = $\frac{1}{4}$"

PUSH STICK PATTERN

3. Round over the inside and outside edges of the handle.

Mount a piloted ¼″ quarter-round bit in your router. Round over the inside and outside edges of the handle, where you'll grip the push shoe. (See Figure 2.) Do not round the edges where the handle attaches to the base.

4. Rout a dado in the base for the heel.

Mount a ¾″ straight bit in your router, and cut a shallow dado near the back edge of the base. (See Figure 3.) Don't attach the heel just yet. If you're making a push block, omit this step.

5. Attach the handle to the base.

Glue the handle to the base and let the glue set up. Then reinforce the glue joints with flathead wood screws, as shown in the working drawings. You must countersink the screws, so they won't mar the wood when you're using the tool. You must also drive the screws at the same angle as the handle.

TIP This angle, by the way, is to help keep your hand out of harm's way. If you're using the push shoe to guide a piece of wood past a blade or cutter, turn the shoe so that the handle is leaning *away* from the danger zone.

Figure 2.   *Round over the handle of the push shoe.*

Figure 3.   *With a straight bit, cut a shallow dado in the base to hold the heel.*

Glue the heel in the dado you made for it. Let the glue cure for at least 24 hours before using the tool.

6. **Glue the heel to the base.**

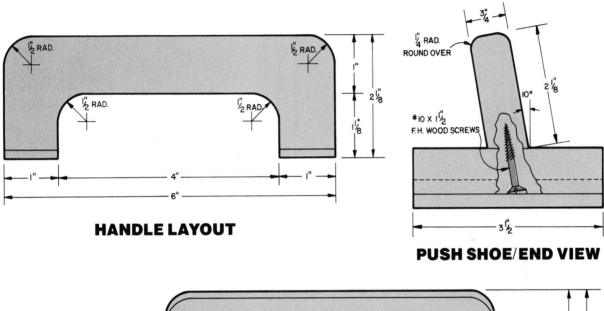

HANDLE LAYOUT

PUSH SHOE/END VIEW

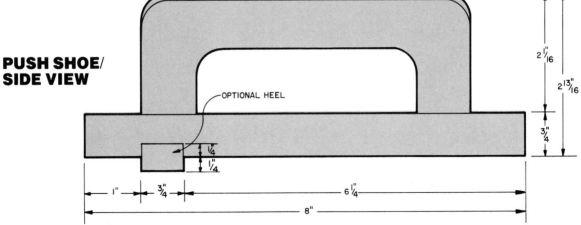

PUSH SHOE/ SIDE VIEW

Pedestal Table

The pedestal table is a classic American furniture design. For almost three centuries, these versatile, three-legged tables have served many different purposes. Traditionally, each leg was joined to the pedestal with a dovetail tenon and a dovetail slot. Before the invention of the router, these dovetails had to be made by hand. But the router, with a dovetail bit, makes the work much easier.

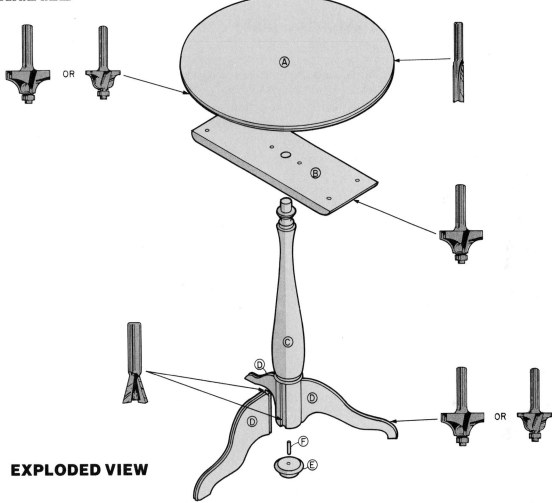

EXPLODED VIEW

Tools and Materials

Wooden parts:

A.	Top	17″ dia. x ¾″
B.	Brace	¾″ x 5″ x 14½″
C.	Pedestal	3″ dia. x 20⅛″
D.	Legs (3)	¾″ x 5″ x 15″
E.	Cap	2½″ dia. x 1¼″
F.	Dowel	⅜″ dia. x 1½″

Hardware:

■ #8 x 1¼″ Roundhead wood screws and flat washers (6)

Router bits needed:

■ ½″ Dovetail bit
■ ½″ Straight bit
■ Piloted edge-shaping bit, such as an ogee or quarter-round bit
■ ½″ Piloted quarter-round bit

Making the Table

1. Turn the pedestal and the cap.

Turn the pedestal and the cap on a lathe, as one piece. The bottom cylinder should be $\frac{1}{8}''$-$\frac{1}{4}''$ longer than shown, to give yourself room to cut the cap from the pedestal. Finish sand the pedestal on the lathe, then separate the cap from the pedestal with a parting tool.

2. Cut the shape of the legs.

Lay out the shape of the legs on the leg stock, paying careful attention to the grain direction. Cut the legs on a bandsaw.

3. Cut the dovetail slots in the leg.

Mount the pedestal in the spindle-holding jig. (*Important:* The lower end of the pedestal must be centered in the jig, and the axis of the spindle must be parallel to the table.) Mount a $\frac{1}{2}''$ dovetail bit in your router, and adjust its height to cut a slot $\frac{7}{16}''$ deep in the bottom end of the pedestal. Clamp a stop block to the fence to stop the cut.

Carefully feed the spindle into the bit, until the jig hits the stop block. (See Figure 1.) Turn the spindle 120° and repeat. Turn the spindle another 120°, and repeat again. When you're done, you should have cut three identical dovetail slots in the spindle.

TIP You can also perform this operation on a Sears/Craftsman 'Router-Crafter', if you have this accessory.

Figure 1. *Mount the pedestal in the spindle-holding jig and cut three dovetail slots, $\frac{7}{16}''$ deep and $3\frac{1}{4}''$ long.*

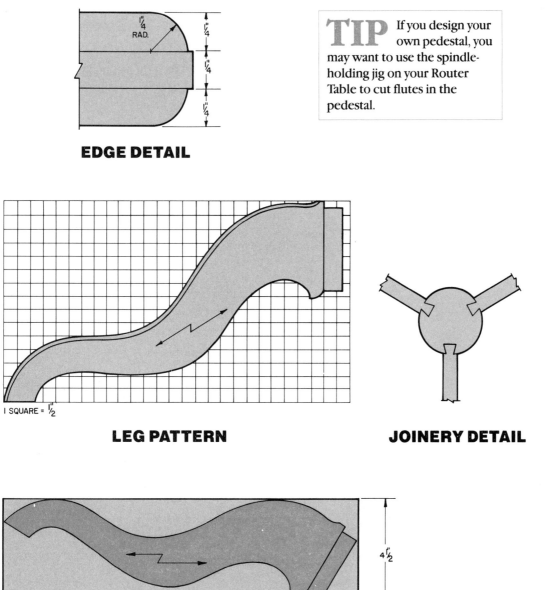

EDGE DETAIL

1/4" RAD.

1/4"
1/4"
1/4"

1 SQUARE = 1/2"

LEG PATTERN

JOINERY DETAIL

> **TIP** If you design your own pedestal, you may want to use the spindle-holding jig on your Router Table to cut flutes in the pedestal.

15 1/4"

4 1/2"

LEG LAYOUT

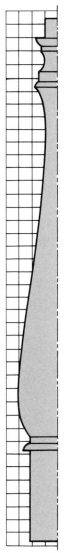

1 SQUARE = 1/2"

POST PATTERN

4. **Cut flats above the slots.**

Draw two parallel lines on either side of each slot, ¾" apart, 3½" long, with the slots centered between them. Using these lines as guides, cut three ¾" x 3½" 'flats' on the lower part of the pedestal with a sharp chisel. (See Figure 2.)

5. **Cut dovetail tenons in the legs.**

Readjust the height of the dovetail bit ⅜" above the worktable, and reposition the fence ⁹⁄₁₆" away from the edge of the bit. Select a piece of ¾" thick scrap wood and pass it between the bit and the table. Turn the wood over and make another pass. The bit will cut a ⅜" long dovetail tenon in the wood. Test fit this tenon to the slots. If it doesn't fit properly, readjust the position of the router bit or the fence, as needed.

After you've properly positioned the fence and the bit, cut the dovetail tenons on all three legs. (See Figure 3.) Pare down the upper ends of the tenons with a chisel to match the rounded ends of the slots. *Important:* If you didn't properly center the pedestal in the jig when you cut the slots, you may have to readjust the fence and the bit for *each slot* to get a proper fit.

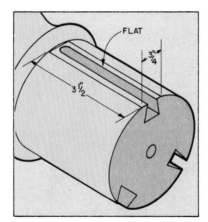

Figure 2. *With a chisel, cut ¾" wide, 3½" 'flats' above each slot. The slots must be precisely centered in the flats.*

Figure 3. *Cut dovetail tenons in the legs by passing the legs between the dovetail bit and the fence. It takes two passes to cut one tenon.*

Cut the round tabletop with the circle-cutting pin and a ½″ straight bit. (See Figure 4.) Take small bites, cutting just ⅛″ deeper with each revolution.

6. **Cut out the tabletop.**

Figure 4.   *Use the circle-cutting pin and a straight bit to cut the top. The top should be 17″ in diameter.*

Mount a ¼″ quarter-round bit, or other piloted edge shaping bit, in the router, and shape the edge of the table and the top edge of the legs. (See Figures 5 and 6.)

7. **Shape the edge of the tabletop and the legs.**

Figure 5.   *Shape the edge of the top. We show a ¼″ quarter-round bit, but you can use whatever suits your fancy.*

Figure 6.   *Shape the top edge of the legs to match the edge of the top.*

8. Make the brace.

Round the bottom edge of the ends of the brace, using a ½″ quarter-round bit. Drill a 1″ hole in the center and test-fit the brace to the top of the pedestal.

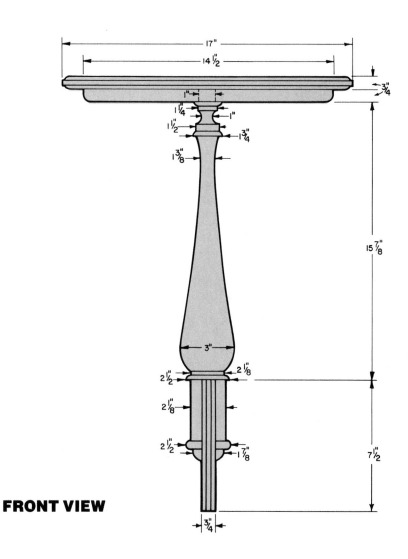

FRONT VIEW

Finish sand all parts. Attach the brace to the tabletop with screws. (*Do not* glue the brace to the top.) Glue the legs to the pedestal and the pedestal to the brace. When the glue dries, apply a finish to the completed table.

9. **Assemble and finish the table.**

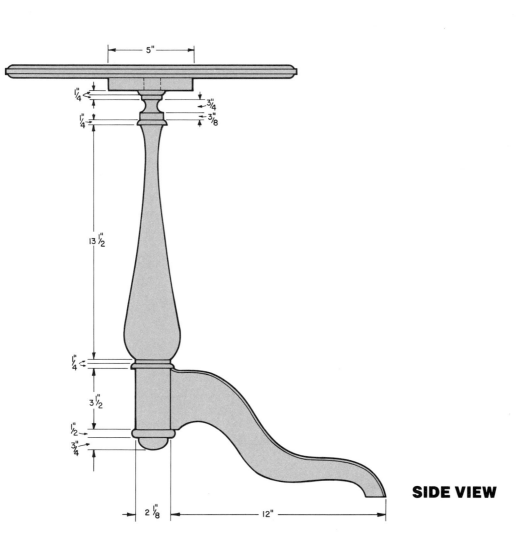

SIDE VIEW

Wine Rack

Good wine isn't just meant to be enjoyed; it's meant to be displayed. A display of good wines heightens the anticipation, and this, in turn, enhances the enjoyment. Your guests ponder over which wine you'll choose: The Chateau Lafite-Rothschild or the Boone's Farm? The Dom Perignon or the Thunderbird? The anticipation prepares the palate.

Of course, in order to display good wines, you need a good wine rack. Here's a simple-but-elegant rack you can put together in just a few hours. As designed, it will hold eight bottles. If you want it to hold more, just make the sides taller or the rails longer.

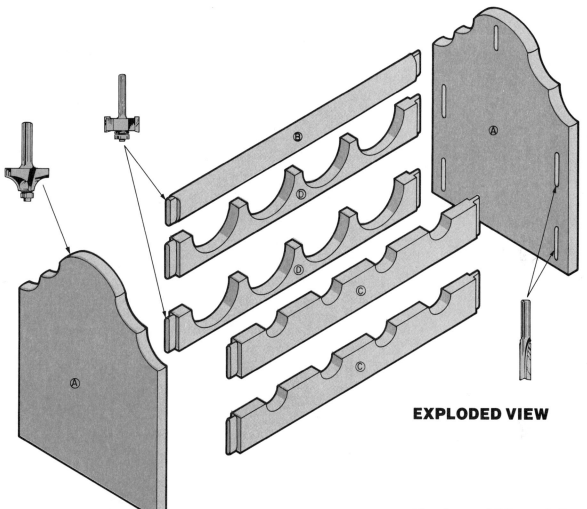

EXPLODED VIEW

Tools and Materials

Wooden parts:

A. Sides (2)   ¾″ x 10″ x 13″
B. Top rail   ¾″ x 1½″ x 18½″
C. Front
    rails (2)   ¾″ x 2⅛″ x 18½″
D. Back
    Rails (2)   ¾″ x 2⅛″ x 18½″

Router bits needed:

- ¼″ Straight bit
-   Rabbeting bit
- ¼″ Quarter-round bit

Making the Wine Rack

1. **Cut the parts to size.**

Cut the sides and top rail to size. To make the front and back rails, cut two pieces of wood 4⅜″ wide and 18½″ long. Later, you'll split these two pieces to make the rails.

2. **Cut the mortises and tenons.**

Use a rabbeting bit to cut the tenons in the ends of the top, front, and back rail stock. Make the matching mortises in the sides with a straight bit, using a straightedge as a guide. (See Figure 1.) With a rasp, round the edges of the tenons to fit the mortises.

3. **Shape the rails and sides.**

With a circle cutter, cut 3½″ holes down the middle of one rail board, and 1½″ holes down the other. Rip the rail stock in two. (See Figure 2.) You'll end up with four rails—two front, two back—each 2⅛″ wide. Cut the shape of the sides on a bandsaw, then round the edges of the sides and the top rail with a ¼″ piloted quarter-round bit.

Figure 1. *Cut mortises in the sides with a straight bit. Use a straightedge, clamped to the side stock, as a guide.*

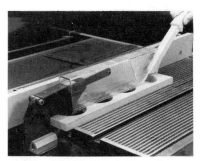

Figure 2. *To shape the rails, cut holes down the middle of the stock, then rip the stock in two.*

4. **Assemble and finish the wine rack.**

Finish sand all parts, then assemble the wine rack with glue. After the glue has cured, do any necessary touchup sanding, and apply a finish to the completed rack.

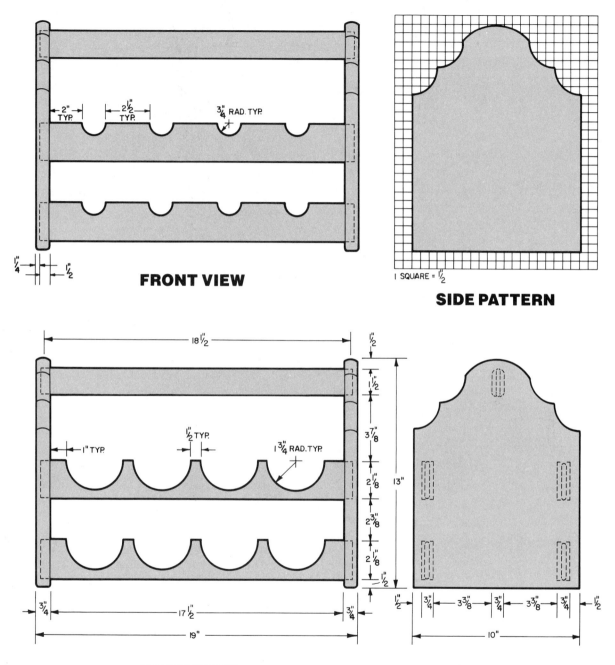

FRONT VIEW

SIDE PATTERN

1 SQUARE = 1½"

BACK VIEW

SIDE VIEW

Breadbox

There is something magical about a tambour door. As you open it, it seems to disappear into some secret compartment in the project, only to reappear when you close the door again. In fact, that's exactly what happens! The shelf in this breadbox isn't attached to the back. As you open the door, the tambours roll out of sight, behind the shelf.

Tambour doors can be easily made with a router. The router is used not only to shape the tambours, but to cut the grooves they ride in. All it takes is a simple template and a guide bushing.

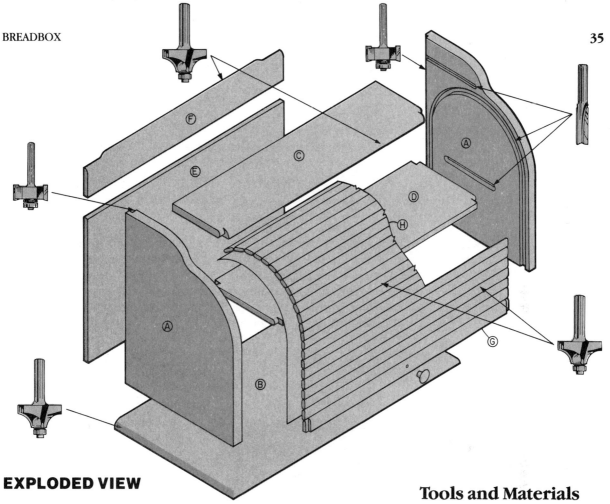

EXPLODED VIEW

Tools and Materials

Wooden parts:

A. Sides (2)	¾″ x 9¾″ x 13¼″
B. Bottom	¾″ x 10¾″ x 21″
C. Top shelf	⅜″ x 4½″ x 18½″
D. Middle shelf	⅜″ x 6¼″ x 18½″
E. Back	¼″ x 11³/₁₆″ x 18½″
F. Backstop	¼″ x 2¹/₁₆″ x 18½″
G. Lead tambour	½″ x 1⅛″ x 18⅜″
H. Tambours (17)	⁵/₁₆″ x ¾″ x 18⅜″

Hardware:

- #8 x 1¼″ Flathead wood screws (6)
- 1¼″ dia. Drawer pull and mounting screw
- 13″ x 17″ Piece of unbleached muslin

Router bits needed:

- ¼″ Quarter-round bit
- ⅜″ Straight bit
- ½″ Quarter-round bit

Router accessories needed:

- ⅝″ Guide bushing

Making the Breadbox

1. Cut all parts to the proper size and shape.

To make the breadbox, you'll need woods of several different thicknesses—¼″, ⅜″, ½″, and ¾″. (Plane the wood yourself, or take it to a cabinet shop or lumberyard and have them do it for you.) Cut out all the parts of the breadbox *except* the tambours. (You *can* cut the lead tambour.) Notch the top and middle shelves, using a bandsaw or saber saw.

2. Cut the joinery in the sides and the lead tambour.

The joinery in the sides—rabbet, blind dado, and double-blind dado—are all made with the router, but each joint uses a different bit or technique. To make the rabbet for the back and backstop, use a rabbeting bit. (See Figure 1.) Also use this bit to cut the rabbets in the ends of the lead tambour. To make the blind dado (for the top shelf), use a ⅜″ straight bit. Using the fence as a guide and a stop block to stop the cut, make a 4″ long dado. (See Figure 2.) To make the double-blind dado (for the middle shelf), drill ⅜″ holes, ⅜″ deep to mark the beginning and end of the dadoes. Put the router in position with the bit in one of the stop holes. Using a straightedge as a guide, cut to the other hole. (See Figure 3.)

Figure 1. *Cut the rabbets in the back edges of the sides with a rabbeting bit.*

Figure 2. *Use a straight bit to cut the blind dado for the top shelf. Use the fence as a guide, and a stop block to stop the dado when it reaches 4″ in length.*

Figure 3. *To make a double-blind dado for the middle shelf, drill stop holes at either end of the dado. Using a straightedge as a guide, rout the dado between the holes with a straight bit.*

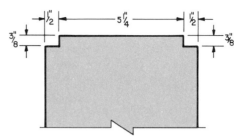

BACK-TO-SIDE JOINERY DETAIL

SHELF JOINERY DETAIL

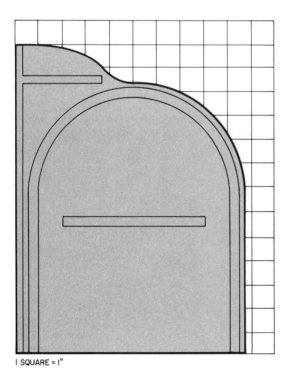

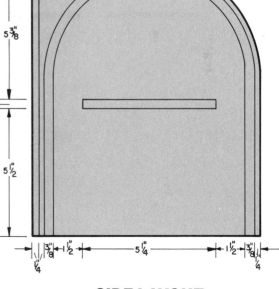

SIDE VIEW PATTERN

SIDE LAYOUT

3. Make the groove for the tambours.

Make a template to help cut the tambour grooves, as shown. This template should be ¼"- ½" thick, depending on how long your ⅝" guide bushing is. Clamp the template to the side stock and cut the groove with a ⅜" straight bit, keeping the guide bushing pressed up against the edge of the template. (See Figure 4.)

4. Cut the shapes of the sides and the backstop.

When you've cut all the joinery in the sides, cut the shapes of the sides and the backstop on a bandsaw. Sand the edges smooth.

5. Make the tambours.

Cut an 18⅜" length of clear, straight ¾" thick stock. Round the top and bottom corner of one long edge, then rip a ⁵/₁₆" thick tambour from the rounded edge. (See Figure 5.) Repeat until you have seventeen tambours. Also, round the top outside edge of the lead tambour.

Figure 4. Use a template to guide the router when you cut the grooves for the tambours. Attach a ⅝" guide bushing to the router, and cut the groove with a straight bit.

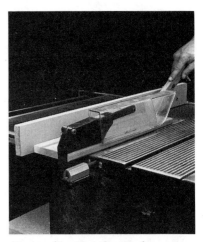

Figure 5. Cut the tambours from the stock after you've rounded the edges. Never try to shape thin stock.

Spread out some waxed paper on your workbench, then tack the muslin to this waxed paper. Spread glue evenly over the muslin, then arrange the tambours, edge to edge. The tambours should stick out on either side of the cloth. (The ends of the tambours must *not* be glued to the muslin.) Put a piece of plywood over the tambours and weight it down. When the glue dries, trim the top and bottom edges of the muslin flush with the tambours.

6. **Glue the tambours to the muslin.**

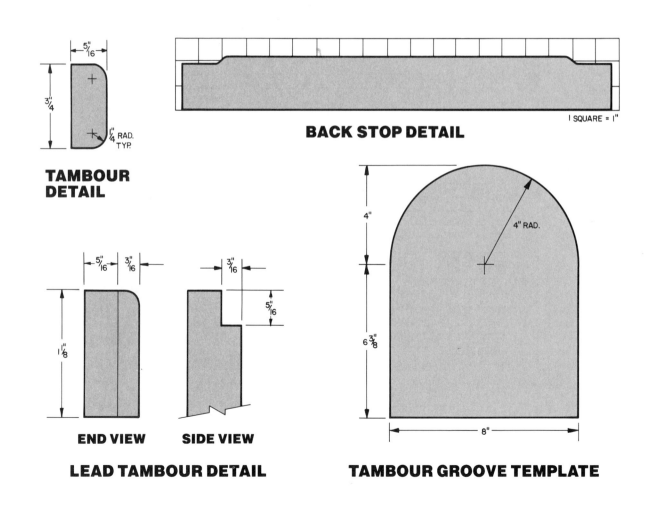

TAMBOUR DETAIL

BACK STOP DETAIL

I SQUARE = I"

END VIEW **SIDE VIEW**

LEAD TAMBOUR DETAIL

TAMBOUR GROOVE TEMPLATE

7. Shape the top shelf, backstop, and bottom.

With a ¼″ piloted quarter-round bit, round over the front and side edges of the bottom, the top edge of the backstop, and the front edge of the top shelf.

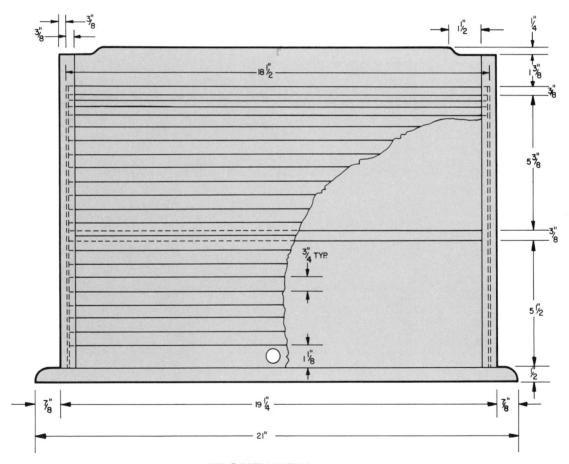

FRONT VIEW

Finish sand all pieces of the breadbox. Assemble the box with glue and brads, then slide the tambour door into place. Attach the bottom to the box assembly with screws. Finally, apply a non-building finish, such as tung oil or Danish oil, and attach a drawer pull to the lead tambour.

8. **Assemble and finish the box.**

TIP Apply some paraffin wax inside the grooves for the tambours to help them slide easily.

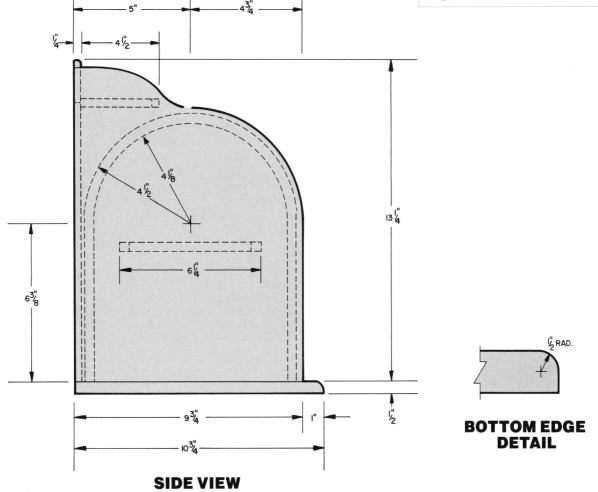

SIDE VIEW

BOTTOM EDGE DETAIL

Shaving Mirror

Shaving mirrors were once a necessary accessory in a man's dressing room, where they were used to hold all the equipment needed for a daily shave. Today, they are used by both men and women for a variety of functions. In the bedroom or the bathroom, they help to organize loose change, keys, jewelry, and dozens of tiny items that would otherwise be spread out over the top of the dresser or cabinet.

The shaving mirror shown here can be easily adapted to suit your needs and accommodate whatever you need it to hold. Change the dimensions of the base to hold a bigger drawer—or several drawers, if you need them. If you enlarge the base, enlarge the mirror too, to keep both elements in the proper proportion to one another.

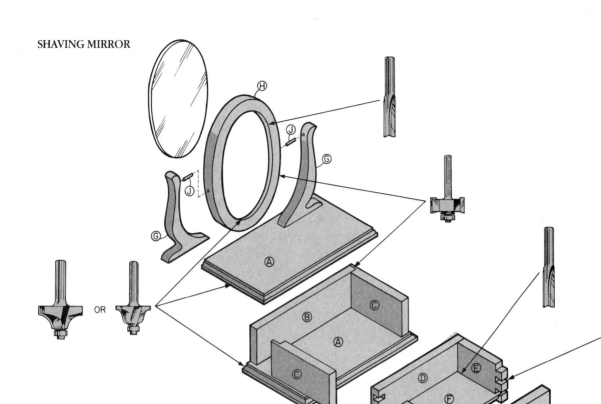

EXPLODED VIEW

Tools and Materials

Wooden parts:

A. Base/
top (2)   ¾″ x 6½″ x 11½″
B. Back   ½″ x 3″ x 10¼″
C. Sides (2)   ½″ x 3″ x 6″
D. Drawer front/
back (2)   ½″ x 2¹⁵⁄₁₆″ x 9⁷⁄₁₆″
E. Drawer
sides (2)   ¾″ x 2¹⁵⁄₁₆″ x 5¼″
F. Drawer
bottom   ¼″ x 5″ x 8¹⁵⁄₁₆″
G. Mirror
supports (2)   ¾″ x 4⅜″ x 6⅝″

H. Mirror frame   ¾″ x 7¼″ x 9¼″
J. Dowels (2)   ¼″ dia. x ¾″

Hardware:

- #8 x 1¼″ Flathead wood screws (4)
- ¾″ Drawer pulls (2)
- 6″ x 8″ Oval mirror
- Glazing points (4)

Router bits needed:

- Rabbeting bit
- ¼″ and ⅜″ Straight bits
- ½″ Dovetail bit
- Piloted edge-shaping bit, such as quarter-round or ogee

Router accessories needed:

- ⁷⁄₁₆″ and ⅝″ Guide collars
- Dovetail template

Making the Shaving Mirror

1. Make the oval mirror frame.

Make an oval template to cut out the inside of the mirror frame. The opening in the template should be 5½″ by 7½″—¼″ larger than the oval the template will help to cut. Mount a ⅝″ guide collar and a ⅜″ straight bit in your router. Attach the mirror frame stock to the template, then drill a starting hole. Starting in that hole, cut the oval with the router, keeping the guide pressed up against the template. (See Figure 1.) Switch to a rabbeting bit and cut the rabbet in the back of the frame stock. (See Figure 2.) *Do not* cut the frame free from the stock—yet.

2. Cut the joinery in the sides.

Cut all pieces for the box to size. With the rabbeting bit still in the router, cut the rabbets in the sides.

3. Shape the top, base, and mirror frame.

Mount a piloted edge-shaping bit in the router, such as a quarter-round or an ogee. Shape the side and front edges of the base and the top. Also shape the inside of the mirror frame. (See Figure 3.)

Figure 1. *With the help of a template, cut the oval opening for the mirror. Cut this opening in several passes, biting just ⅛″-¼″ deeper with each pass.*

Figure 2. *With a rabbeting bit, cut the rabbet in the back of the mirror frame stock.*

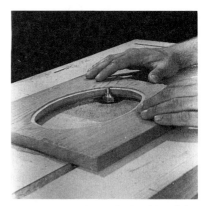

Figure 3. *Shape the inside of the mirror frame stock with a piloted edge-shaping bit. Only* after *the frame opening has been cut, rabbeted, and shaped, should you cut the frame free from the stock.*

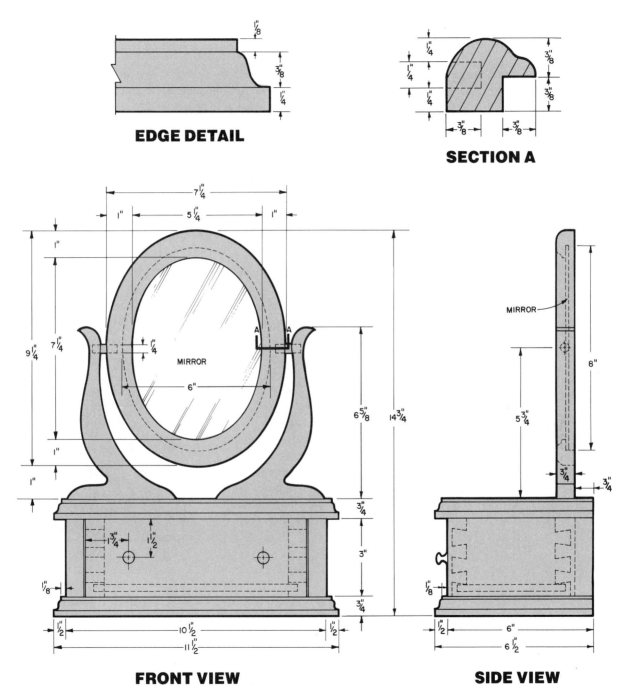

EDGE DETAIL

SECTION A

MIRROR

MIRROR

FRONT VIEW

SIDE VIEW

4. Cut the mirror frame and the support arms.

With a bandsaw, cut the mirror frame free from the stock. Trace the shape of the arms on clear, straight-grained stock, and cut the arms. Drill holes for the supporting dowels in the frame and the arms.

5. Make the drawer.

With a ½″ bit, a ⁷⁄₁₆″ guide collar, and a dovetail template, cut the half-blind dovetails that join the drawer front, back, and sides. (See Figure 4.) Since the stock is just ½″ thick and most dovetail templates are made for ¾″ stock, you will probably have to shim the stock or adjust the template to compensate. Cut the grooves for the drawer bottom with a ¼″ straight bit.

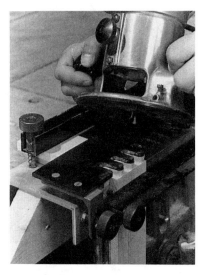

Figure 4.   With a dovetail template, cut the half-blind dove-tails that hold the drawer parts together. Since the drawer stock is ½″ thick and most templates are made for ¾″ stock, you may have to shim the stock or adjust the template.

6. Assemble and finish the shaving mirror.

Finish sand all parts. Assemble the drawer parts with glue. Insert the dowels in the arms and the frame, then glue the mirror assembly to the top. Reinforce the glue joint between the top and the arms with screws. Then glue the top, base, sides, and back together. Do any necessary touchup sanding, then apply a finish. Finally, mount the mirror in the frame and attach pulls to the drawer.

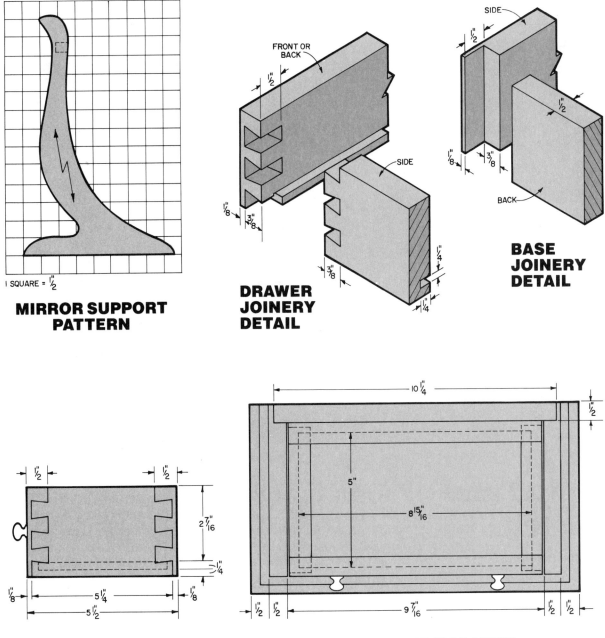

I SQUARE = 1½"

MIRROR SUPPORT PATTERN

FRONT OR BACK

½"

⅛"
3/8"

DRAWER JOINERY DETAIL

SIDE

3/8"

1/4"

1/4"

SIDE

½"

½"

⅛"
3/8"

BACK

BASE JOINERY DETAIL

½"

½"

2 7/16"

1/4"

⅛"

5 1/4"

⅛"

5 1/2"

DRAWER/SIDE VIEW

10 1/4"

½"

5"

8 15/16"

½" ½"

9 7/16"

½" ½"

BASE/DRAWER/TOP VIEW

Outlet and Switch Covers

Sometimes, the little touches can add so much to the decor of a room. Hand-crafted wooden covers on your electrical switches and outlets, for example, convey a feeling of elegance and warmth that is out of proportion to their small size. You'd think that you'd hardly notice them. But the truth is that you look at them every time you turn a light on or off.

Making outlet and switch covers can be a bit tricky. They have to fit precisely, and this takes a lot of hand work. But you can cut down on the amount of work you need to do by using your Router Table to make the covers. By using the pin routing technique, you can make just one cover, then duplicate it quickly and precisely, as many times as you need.

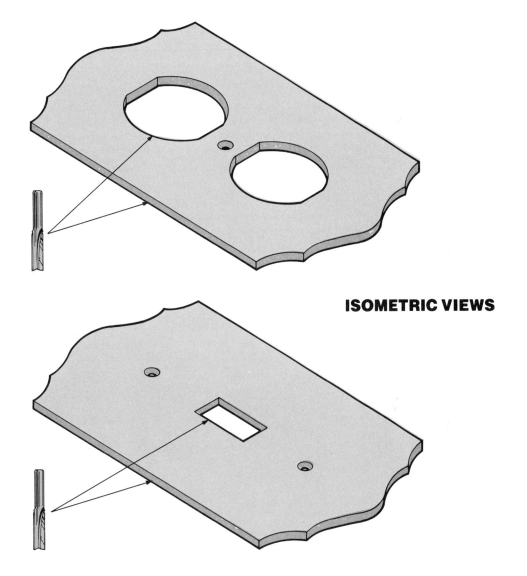

ISOMETRIC VIEWS

Tools and Materials

Wooden parts:

A. Outlet/switch
   cover   ⅛″ x 2⅞″ x 4⅜″

Router bits needed:

- ¼″ Straight bit

Making the Covers

1. **Make pin routing templates for the covers.**

Select hardwood or particle board stock to make the templates. Sink cutouts—scraps of particle boards covered with plastic laminate—work best. You can purchase them at most building supply centers. Carefully draw or trace the outline of the switch and outlet covers on the template blanks. Be careful that you get the opening *exact*. (See Figure 1.)

Mount a ¼″ straight bit in your router, and carefully rout a ¼″ groove, outlining the covers. You can use a straightedge as a guide to rout the straight parts of the groove, but you'll have to rout the curves freehand. (See Figures 2 and 3.) Take it slow, take very shallow bites, and follow the lines as closely as you can. With a little practice, you'll be surprised at how accurately you can make freehand curves.

When you've finished routing the templates, compare them to your original patterns. Pay particular attention to the grooves outlining the openings for the outlets or the switch toggle. They must be accurate.

Figure 1. *Trace the shape of the cover on the template blank. To make sure you get the openings exact, use a commercial outlet or switch cover as a pattern.*

Figure 2. *When routing the template, use a straightedge to guide the straight cuts.*

On your bandsaw, resaw hardwood stock to $3/16''$ thick. Then take the stock down to $1/8''$ thick with a belt sander or planer. If you don't want to go through this trouble, you can also buy $1/8''$ thick wood from companies who supply wood to musical instrument makers.

2. Prepare ⅛″ thick stock for the covers.

Set up your Router Table for pin routing. Mount a $1/4''$ straight bit in the router, and a $1/4''$ pin in the arm. Attach the stock to the bottom of the template, tacking it in place with small brads. Make sure you place these brads where you will later drill holes for the mounting screws. That way, you won't see the brad holes in the finished covers.

Rout the outside shapes of the covers, letting the pin ride in the grooves of the template. (See Figure 4.) Since the stock is only $1/8''$ thick, you'll probably be able to cut the covers in one pass. Remove the scrap, then rout the openings.

3. Cut the shapes of the covers.

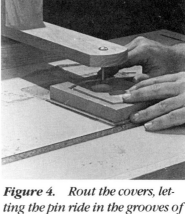

Figure 3. *Carefully rout the curves freehand. And remember to make a 'lead-in' groove to help rout the outside shapes.*

Figure 4. *Rout the covers, letting the pin ride in the grooves of the template.*

4. **Finish shaping the covers.**

The router bit won't cut the square corners you need in the openings; you'll have to do a little hand work to make these corners. Remove the stock from the template and mount them in a vise. Square the corners with a triangular file. (See Figure 5.)

5. **Drill the mounting holes.**

Carefully mark the location of the holes for the mounting screw on the covers. Countersink the hole, then drill a pilot hole for the shank of the screw. If you use decorative roundhead screws, you can forget the countersink.

6. **Finish the covers.**

Finish sand the covers, and apply a stain or a finish. The finish should match the other wood trim or moldings in the room where you intend to use the covers.

Figure 5.   *Where needed, square the corners of the openings with a triangular file.*

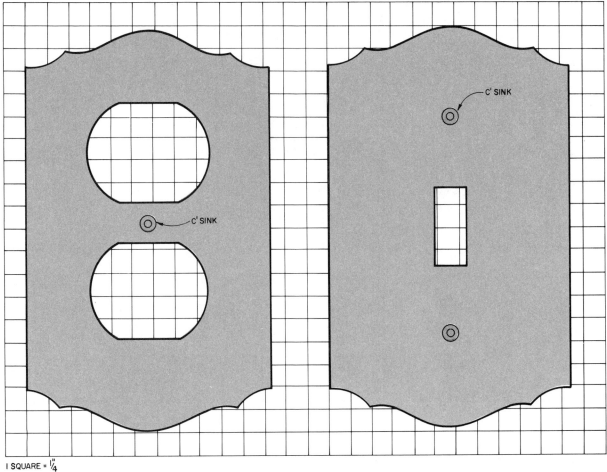

I SQUARE = 1/4"

OUTLET PATTERN   **SWITCH PATTERN**

Foot Stool

Every room needs a foot stool. It's an incredibly useful piece of furniture: You use it to prop your feet up while you're reading, to reach things you've stored on the top shelf at the back of a closet, even as a seat or a makeshift table for a child.

This foot stool seems more delicate than most; but, in fact, it is quite sturdy. The cabriole legs give it a light and airy look, but the mortise-and-tenon joinery adds strength. The foot stool is actually built with the same joints that you would expect to find in a large table. Because of this, it's sturdy enough to endure any task you may want to put it to.

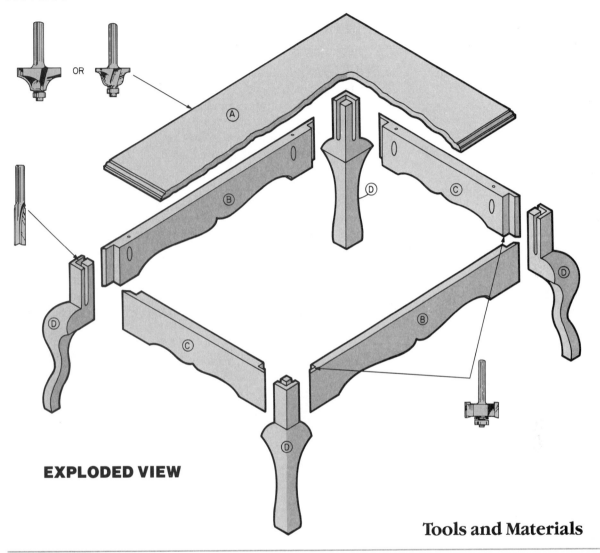

EXPLODED VIEW

Tools and Materials

Wooden parts:

A. Top   ¾″ x 13½″ x 19½″
B. Front/Back
   aprons (2)   ¾″ x 2½″ x 17″
C. Side
   aprons (2)   ¾″ x 2½″ x 11¼″
D. Legs (4)   2″ x 2″ x 9¼″

Hardware:

■ #8 x 1¼″ Flathead wood
screws (8)

Router bits needed:

■ ¼″ Straight bit
■ Rabbeting bit
■ Edge shaping bit, such as a
quarter-round or ogee bit

OR

Making the Stool

1. **Cut the mortises in the leg stock.**

It's best to cut the joinery in the legs *before* you shape them. Cut stock for the legs, and mount a ¼″ straight bit in the router. Position the fence of the Router Table ½″ away from the edge of the bit, and attach a stop block to the fence to stop the mortise when it's 2½″ long. Using the fence as a guide, cut a long mortise on each of the inside faces of each leg—eight mortises, altogether. (See Figure 1.)

2. **Cut the joinery in the aprons.**

Cut the stock for the aprons, and miter the ends at 45°. Using a rabbeting bit, cut tenons in the ends of the apron stock. (See Figure 2.) After you've made the tenons, drill screw pockets to attach the top to the aprons. Tilt the table of your drill press 15° off horizontal, and clamp a scrap of 2 x 4 to the table to serve as a temporary fence. Drill the larger 'pocket' holes first, then the smaller pilot holes. (See Figure 3.) The pilot holes should exit in the middle of the top edge of the aprons.

Figure 1. *Cut the mortises in the legs* before *you shape them. Use the fence as a guide, and attach a stop block to stop the cuts.*

Figure 2. *Cut the tenons in the aprons with a rabbeting bit.*

Figure 3. *Drill the screw pockets at 15°. Drill the pockets first, then the pilot hole.*

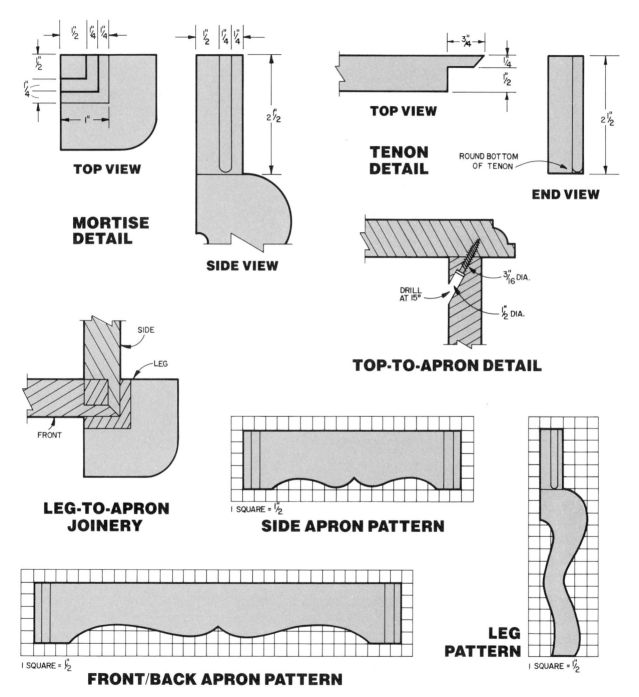

TOP VIEW

MORTISE
DETAIL

SIDE VIEW

TENON
DETAIL

ROUND BOTTOM
OF TENON

END VIEW

DRILL
AT 15°

3/16" DIA.

1/2" DIA.

TOP-TO-APRON DETAIL

SIDE

LEG

FRONT

LEG-TO-APRON
JOINERY

1 SQUARE = 1/2"

SIDE APRON PATTERN

LEG
PATTERN

1 SQUARE = 1/2"

1 SQUARE = 1/2"

FRONT/BACK APRON PATTERN

3. Shape the legs and the aprons.

Trace the pattern of the cabriole legs on the two *inside* faces of each leg. Cut the shape of one face, saving the scrap. Tape the scrap back to the stock, and cut the second face. (See Figure 4.) When you remove the tape, you'll have a cabriole leg. Repeat for all four legs, then sand away any saw marks.

While you're working on the bandsaw, trace the patterns for the aprons onto the apron stock, and cut them out. Sand away any saw marks.

TIP. You can save time by 'pad sawing' and 'pad sanding' the aprons. Just tape the stock for the two side aprons and the two end aprons together. When you're done sawing and sanding, remove the tape.

Figure 4. To shape the cabriole legs, trace the pattern on the two inside faces. Cut one face on the bandsaw; tape the waste back to the stock, and cut the second face.

4. Cut and shape the top.

Cut the top to size, then shape the edge. Use a piloted edge-shaping bit, such as a quarter-round or an ogee. As shown in the working drawings, we used a quarter-round bit, and made two ⅛″ 'steps'.

5. Assemble and finish the stool.

Finish sand all the parts of the stool. Assemble the legs and the aprons with glue. When the glue has cured, do any necessary touchup sanding, and apply a finish to all the parts. Finish both the top and the bottom sides of the top, to keep it from warping. When the finish has dried, screw the top to the apron/leg assembly. *Don't* glue the top in place. Leave it free to expand and contract slightly with changes in humidity.

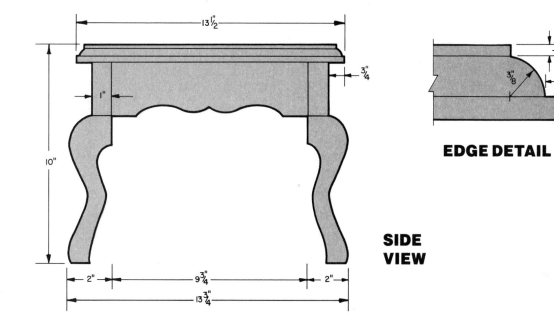

EDGE DETAIL

SIDE VIEW

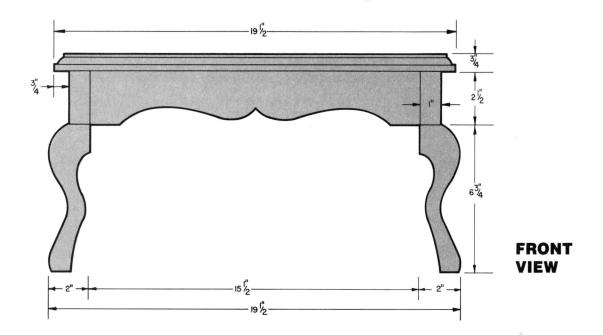

FRONT VIEW

Platters and Cutting Boards

There is nothing better for your cutlery than a wooden cutting board or meat platter. The wood, because it is softer than metal, keeps the knives sharper longer.

However, wood also absorbs the juices from the vegetables, fruits, and meats that it comes in contact with. The left over, soaked-in juice from one type of food can taint the flavor of another, if you're not careful. For this reason, it's wise to keep several different wooden platters around the kitchen, each one to be used to cut or carve just one particular type of food—vegetables, fruits or meats.

Shown here are two cutting boards/platters for all sorts of purposes. One side is flat; the other has grooves to keep juices from running over onto your counter. Select the style that suits your needs and tastes.

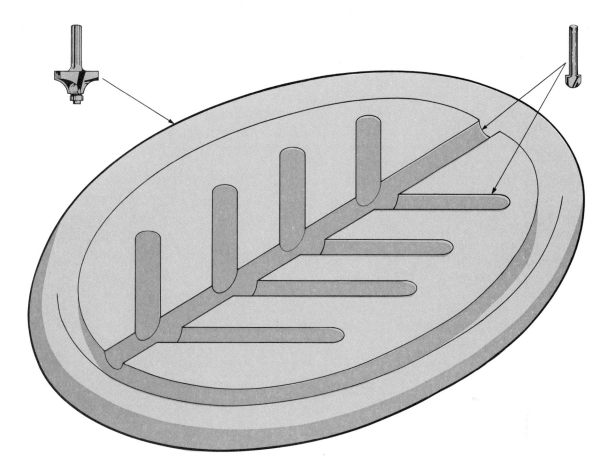

ISOMETRIC VIEW

Tools and Materials

Wooden Parts:

A. Large
   platter   ¾″ x 10¾″ x 16¾″
B. Small platter   ¾″ x 8″ x 11″

Router bits needed:
- ¼″ Piloted quarter-round bit
- ½″ Core-box bit

Router accessories needed:
- ⅝″ Guide collar

Making the Platters

1. Rout the grooves in the platter stock.

It's best to rout the grooves in a platter *first,* before you cut the shape. To rout the grooves, first cut a template out on the bandsaw. Make the template ½″ thick, and ⁵⁄₁₆″ smaller than the circumference of the grooves. Attach a ⅝″ guide bushing to the base of the router, and mount a ½″ core-box bit. Attach the template to the stock with double-faced carpet tape. With the guide pressed firmly against the template, rout the grooves. (See Figure 1.)

> **TIP** If the grooves are symmetrical, you may only need to make *half* a template. Rout half of the grooves, flop the template over, and rout the rest.

Figure 1.   Use a template and a guide bushing to guide the router when cutting the groove in the platter stock. Cut this groove before cutting out the shape of the platter.

2. Shape the platter.

Cut the shape of the platter out on a bandsaw, and sand away the saw marks. With a ¼″ piloted quarter-round bit, round the edges.

3. Finish the platter.

Finish sand all surfaces, and coat the platter with a non-toxic finish such as mineral oil or salad bowl dressing.

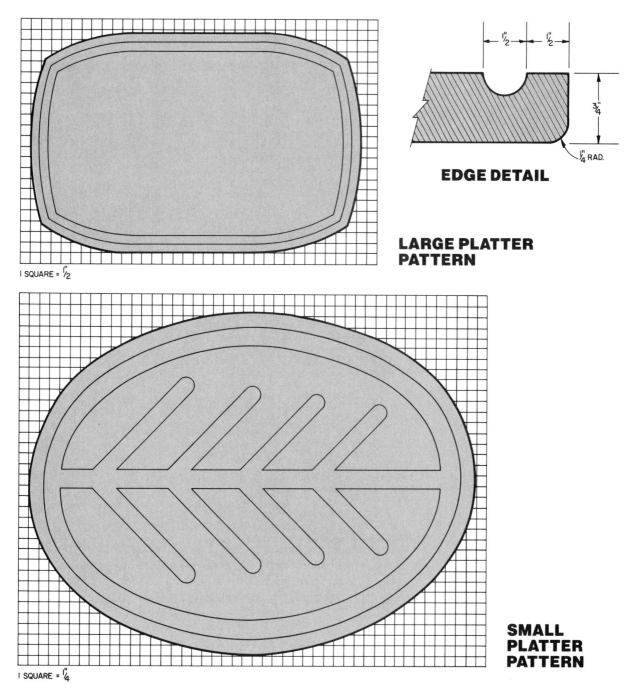

I SQUARE = 1/2"

EDGE DETAIL

LARGE PLATTER PATTERN

1/2" 1/2"

3/4"

1/4" RAD.

SMALL PLATTER PATTERN

I SQUARE = 1/4"

Hand Mirror

"Craftsmanship is the marriage of beauty and utility," wrote a nineteenth-century artisan, J. Geraint Jenkins. If that's true, then this elegant hand mirror is the soul of craftsmanship. It is simple, utilitarian, but its graceful lines are pleasant to hold, wonderful to look at.

It's a simple project to make, too. The mirror is just two pieces of wood, one of them routed to hold a round mirror. Then the wooden parts are laminated together and shaped—something that you can do in an evening.

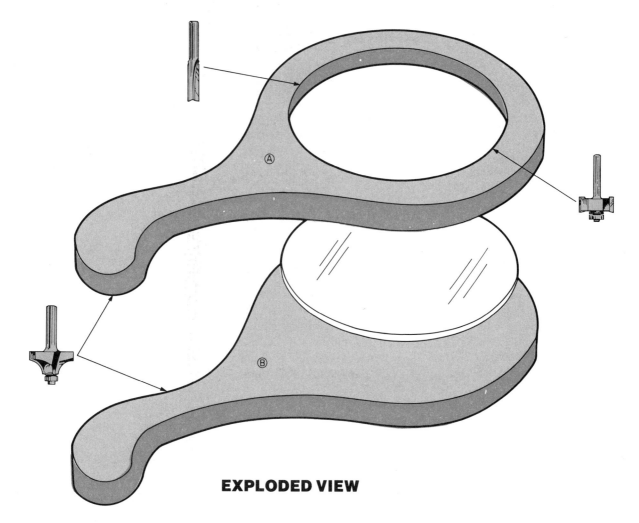

EXPLODED VIEW

Tools and Materials

Wooden parts:

A. Frame   ⅜″ x 6″ x 11½″
B. Back   ⅜″ x 6″ x 11½″

Hardware:

- 5″ dia. Mirror

Router bits needed:

- ½″ Straight bit
- Piloted rabbeting bit
- ¼″ Piloted quarter-round bit

Making the Hand Mirror

1. **Resaw and plane the stock.**

The two wooden halves of the hand mirror are each ⅜″ thick. Since this is not a standard lumber size, you must resaw and plane the wood to the proper thickness. Resaw the wood first on your bandsaw to ⁷⁄₁₆″ thick, then remove the final ¹⁄₁₆″ in the planer. Try to saw both halves from the same board, and mark the edge before you resaw. Refer to the marks when you glue up the hand mirror, and try to put the pieces together the same way they came apart. If you're careful, no one will be able to tell that the mirror is made from two pieces of wood.

Option: You may also want to make the mirror from two *contrasting* pieces of wood, such as walnut and maple, rosewood and cherry, etc. The contrasting colors and grain make a nice effect.

TIP If you don't have a plane to surface the stock, use a belt sander or take it to a local lumberyard. They'll plane it for you for a small fee.

2. **Cut the *inside* of the frame.**

Mount a ½″ straight bit in your router, and mount the circle cutting pin on the Router Table so that it's 2″ away from the axis (or center) of the router bit. Carefully measure and mark the center of the circular opening in the frame stock. Mount the pin so that it pivots at the center mark, and cut a circle out of the frame. (See Figure 1.) Take small bites, removing no more than ⅛″ at a time. When you're finished, the opening should be 4½″ in diameter.

Figure 1. *Using the circle cutting pin, rout the opening for the mirror in the frame blank.*

With a piloted rabbeting bit, cut a ¼″ wide, ⅛″ deep rabbet all the way around the inside edge of the mirror opening in the frame. (See Figure 2.) Check to see that the mirror fits with little slop.

Figure 2.   *Cut a rabbet in the inside edge of the mirror opening to hold the mirror.*

$3.$ **Make a rabbet for the mirror.**

Put the mirror in the frame, and dry clamp (no glue) the back to the frame. Shake the assembly. If the mirror rattles, it's too loose in the rabbets. To tighten it up, cut little pieces of felt and stick them to the back, near the edge of the mirror. Dry clamp the pieces together again. If the mirror still rattles, add another layer of felt.

When the mirror fits snug, glue the two pieces together with the mirror in place. Let the glue cure for at least 24 hours.

$4.$ **Glue the back and the frame together.**

5. **Cut the handle and the outside shape of the frame.**

Trace the pattern for the hand mirror on to the frame/back assembly. Then cut out the shape on a bandsaw or jigsaw. (See Figure 3.) Remove the saw marks from the cut edges with a drum sander.

6. **Round over all the edges.**

Mount a ¼″ piloted quarter-round bit in the router, then round over all edges of the mirror. (See Figure 4.) Be careful to round over the end grains first, then the long grains.

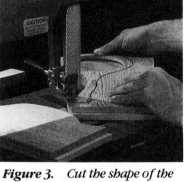

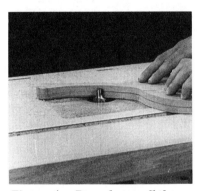

Figure 3. *Cut the shape of the mirror in the back and the frame* only *after* you've assembled the two pieces.

Figure 4. *Round over all the edges of the hand mirror with a quarter-round bit.*

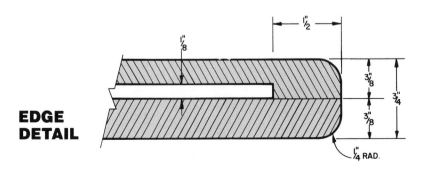

**EDGE
DETAIL**

Finish sand the mirror assembly, then cover the mirror with tape. This tape will protect the mirror while you apply a finish to the wood. All you have to do to clean up any finish which spills over onto the mirror is remove the tape.

7. **Finish the hand mirror.**

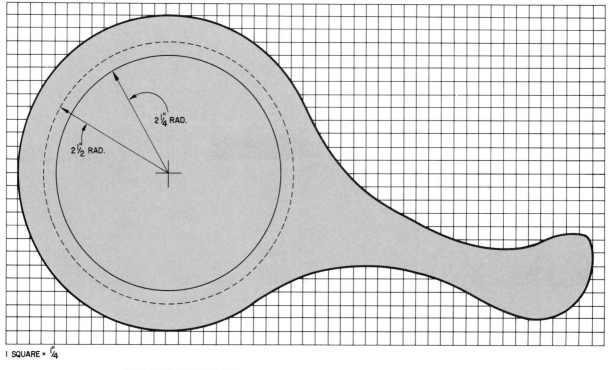

1 SQUARE = 1/4"

HAND MIRROR PATTERN

TIP The pattern you see here is just a suggestion. Design your own hand mirror, if you wish.

Herb Clock

Over the centuries, craftsmen and craftswomen have spent an enormous amount of time devising better ways to tell the time. Here's our small (but timely) contribution to this timeless endeavor: a wooden 'herb' clock for the kitchen.

This project shows another use for the pin arm on your Router Table. When making this clock, the cavities for the herbs in the clock body and the holes in the frame must match up precisely. The easiest way to do this is to make a template and pin rout both parts using this same template.

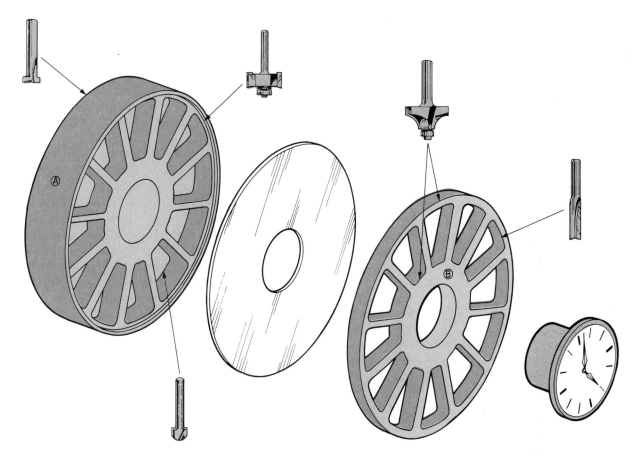

EXPLODED VIEW

Tools and Materials

Wooden parts:

A. Body   9¾″ dia. x 1½″
B. Frame   9¾″ dia. x ½″

Hardware:

■ Clear acrylic plastic, 9¼″ dia. x ⅛″ thick
■ 'Cylinder' clock, battery powered
■ Dried herbs (12 different varieties)

Router bits needed:

■ ¼″ Straight bit
■ Rabbeting bit
■ ½″ Core-box bit
■ ¼″ Piloted quarter-round bit
■ Keyhole bit

Making the Clock

1. Rout the cavities in the body and openings in the frame.

With a jigsaw or a saber saw, make a pin-routing template to rout the cavities in the body and the openings in the frame. Both of these follow the same pattern, so you can use the same template. Cut stock for the body and the frame, but do not cut the round shapes just yet. Wait until after you rout the cavities and the openings.

To make the body—Drill twelve ½″ 'starter holes', ¾″ deep in the body stock, in the middle of the areas where you will rout the cavities. Mount a ½″ core-box bit in the router, and attach the template to the body stock. Put a ½″ pin in the pin arm, and adjust the pin so that it's directly over the bit. Raise the bit to cut ⅛″-¼″ deep, and put the stock in place with the bit in one of the starter holes. Lower the pin so that it will rest against the template, then rout the cavity, using the template to guide your work. Raise the bit and make another pass. Continue until the cavity is ¾″ deep. (See Figures 1 and 2.) Repeat for all twelve cavities.

To make the frame—Use the same procedure as you did for the body, but mount a ¼″ straight bit in the router and drill starter holes all the way through the stock. Rout completely through the frame stock, making twelve openings to match the cavities in the body.

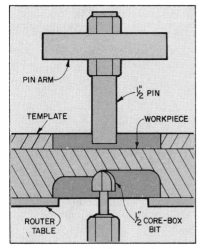

Figure 1.   *Position the pin directly over the bit, then let the pin ride against the template to guide the work.*

Figure 2.   *Rout each cavity in several passes, routing ⅛″-¼″ deeper with each pass. Start each pass in the starter holes.*

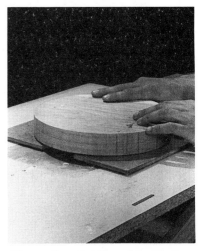

Figure 3.   *Cut the rabbet in the body of the clock with a rabbeting bit. Use the circle cutter to guide the work.*

With a rabbeting bit, rout a ⅛″ deep rabbet all the way around the inside of the body to hold the clear plastic. Use the circle cutter to make this rabbet, spinning the body on the pivot pin. (See Figure 3.) Cut the outermost diameter of the rabbet first, then move the pivot pin in slightly, and cut again. Continue until you can't move the pin any closer to the bit. Rout the last little bit of stock in the center of the body by removing the pin and moving the stock back and forth over the bit. Be careful not to rout the edges!

2. Rout the rabbet in the body.

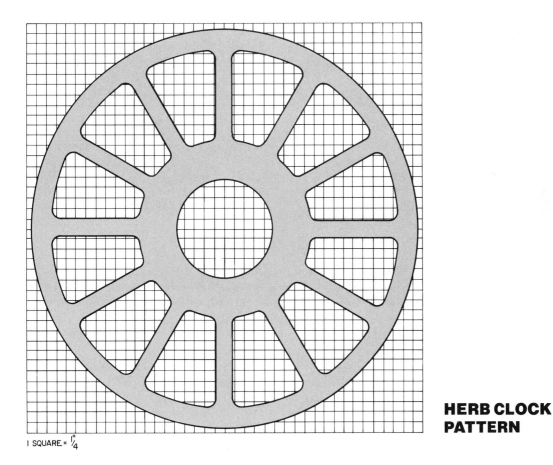

HERB CLOCK PATTERN

1 SQUARE = ¼″

3. **Cut the shapes of the body and frame.**

Temporarily, stick the body and frame together with tape. 'Pad saw' both pieces at the same time, cutting the circular shape with a bandsaw. Sand away any saw marks.

4. **Make a keyhole slot to hang the clock.**

If you wish, rout a keyhole slot in the back of the clock to hang it. Drill a ⅜″ pilot hole, ⅜″ deep in the clock assembly, and adjust the keyhole bit to cut ⅜″ deep. Put the bit in the hole and rout a short slot, about ¾″ long. (See Figure 4.)

Figure 4.   *To hang the clock, cut a keyhole slot in the back with a keyhole bit.*

5. **Round over the edges of the frame.**

With a ¼″ quarter-round bit, round over the edges of the frame. Be very careful during this operation that you don't go too fast and splinter the thin spokes or rim of the frame. Afterwards, finish sand the frame and apply a finish.

6. **Assemble and finish the clock.**

Fill the cavities in the body with different dried herbs. (They *must* be completely dry.) Put the clear plastic in place over the herbs, then glue the frame to the body. After the glue is dry, finish sand the sides of the clock and apply a finish. With a hole saw, cut a large hole in the middle of the assembly for the clock. Then insert the clock in the hole.

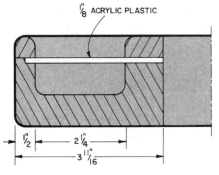

1/8" ACRYLIC PLASTIC

1/2" 2 1/4"

3 11/16"

SECTION A

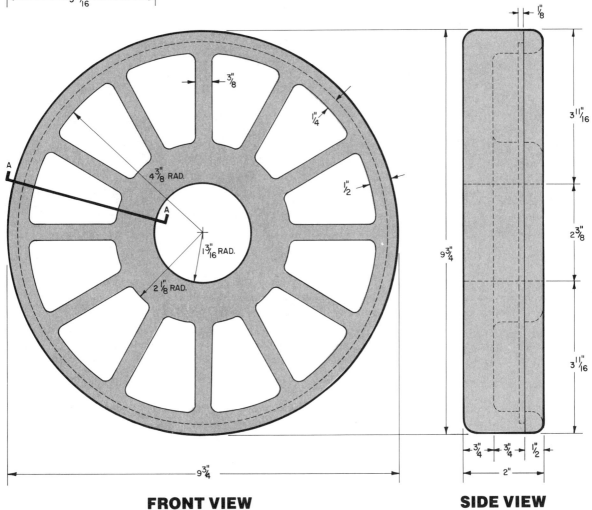

3/8"

1/4"

4 3/8" RAD.

A

A

1 3/16" RAD.

2 1/8" RAD.

1/2"

9 3/4"

9 3/4"

1/8"

3 11/16"

2 3/8"

3/4" 3/4" 1/2"

2"

FRONT VIEW

SIDE VIEW

Coat Rack

Of all the inventions of mankind, the simple peg has to rate as one of the most useful. Think of the storage problems solved, the messes and the confusion avoided over the centuries just because of this simple device. If you don't already have a few pegs hanging around, then you can probably use some. And even if you do, then you may be able to use some more.

Although we call this project a 'coat rack', it is actually much more versatile. The pegs can be used to hang not only coats, but hats, scarves, umbrellas, towels, ties, mugs, anything with a strap or handle. Mount one in your hallway, entranceway, bathroom, closet, kitchen—anywhere you can use hanging storage. We've designed this project so that you can easily change the design to fit in a particular space or fill a particular need. Just lengthen or shorten the rack to include more or fewer pegs than we show here.

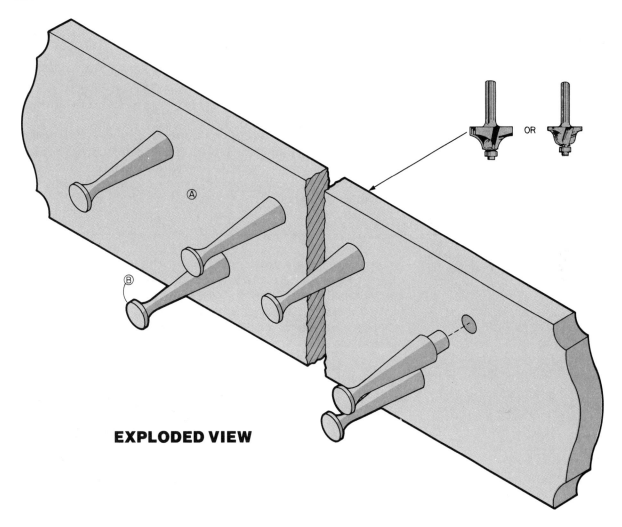

EXPLODED VIEW

OR

Tools and Materials

Wooden parts:

A. Backboard*   ¾" x 5½" x
                  Variable

B. Pegs (Variable)* ⅞" dia. x 3½"

*Use hardwood pegs and a matching
hardwood for the backboard.

Hardware:

■   Molly anchors or expansion
bolts to hang the coat rack

Router bit needed:

■   Piloted edge-shaping bit, such
as an ogee or quarter-round bit

Making the Coat Rack

1. Shape and drill the backboard.

Rip and joint a backboard 5½″ wide, and cut it as long as you need to hold the number of pegs you want. With this design, you'll need an 8″ long board to mount one peg. Each additional peg requires the backboard to be another 2″ longer.

Trace the pattern on both ends of the backboard, then cut the ends on a bandsaw. Sand out the millmarks, then drill holes for the peg shanks in the backboard, where shown. Make the holes approximately ¹/₁₆″ deeper than the length of the mounting shanks, to leave room for the glue. Also, drill wall mounting holes.

2. Shape the edge of the board.

Set up your router as a shaper. Select a *piloted* bit that will cut a shaped edge. There are several available—ogee, bead, or cove. We used a quarter-round bead with two ¹/₁₆″ 'steps'. Cut the end grains first, then the long grains. (See Figures 1 and 2.) By cutting the end grains before the long grains, you can hide the effect of any 'tear-out' at the end of the end-grain cuts.

Figure 1. *Shape the end grains of the backboard first. Don't worry if there is a little tear-out or chipping at the end of the cut.*

Figure 2. *Shape the long grains after you shape the end grains. This will remove the effects of tear-out and chipping.*

3. Assemble and finish the rack.

Finish sand the backboard, then glue the pegs in place. Apply several coats of durable finish. After the finish has dried, attach the rack to the wall.

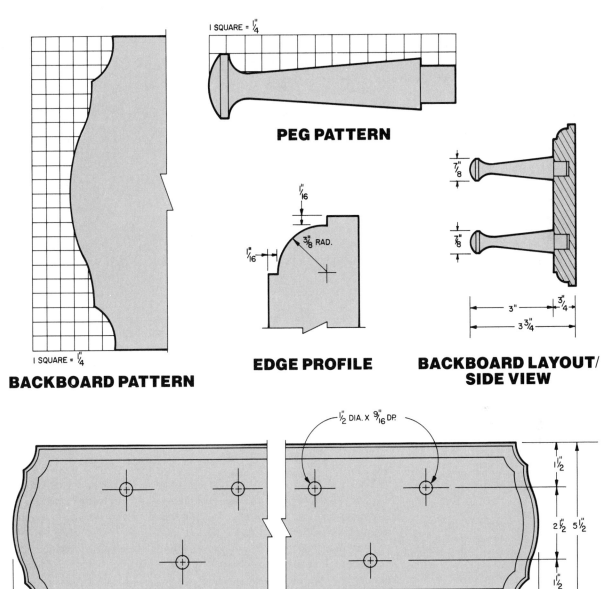

I SQUARE = $\frac{1}{4}$"

PEG PATTERN

I SQUARE = $\frac{1}{4}$"

BACKBOARD PATTERN

$\frac{1}{16}$"

$\frac{3}{8}$" RAD.

$\frac{1}{16}$"

EDGE PROFILE

$\frac{7}{8}$"

$\frac{7}{8}$"

3"

$\frac{3}{4}$"

$3\frac{3}{4}$"

**BACKBOARD LAYOUT/
SIDE VIEW**

$1\frac{1}{2}$" DIA. X $\frac{9}{16}$" DP.

$1\frac{1}{2}$"

$2\frac{1}{2}$"

$5\frac{1}{2}$"

$1\frac{1}{2}$"

4" 2" 2" 2" 2" 4"

VARIABLE

BACKBOARD LAYOUT/FRONT VIEW

Signs

Wooden signs always add a look of handcrafted elegance to a setting. They ought to. There is no other way to make a wooden sign except by hours of tedious hand carving—*unless* you have a router.

The router makes carving wooden signs a breeze. With a router and a pantograph accessory, you can do a professional job with very little experience. We suggest using a pantograph instead of the numerous sign-routing accessories that are available for three reasons. First, with a sign-router, you're usually confined to just one or two styles of numbers. A pantograph lets you cut any style you want. Second, with a pantograph, you can do both raised *and* incised lettering. And finally, a pantograph allows you to add other graphic elements to your signs besides numbers —arrows, animals, the Mona Lisa.

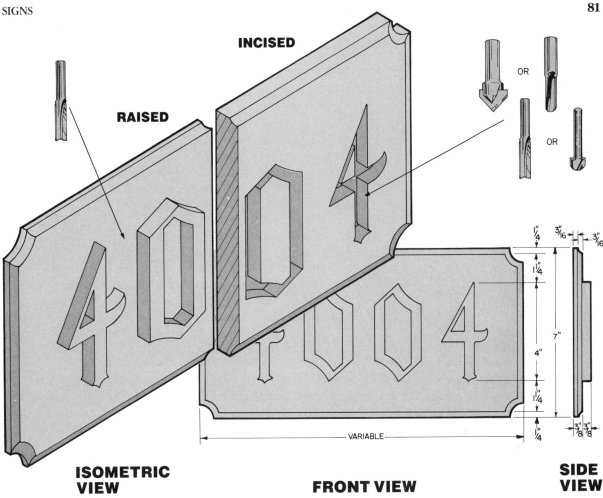

RAISED

INCISED

ISOMETRIC VIEW

FRONT VIEW

SIDE VIEW

Tools and Materials

Wooden parts:

¼" Thick hardboard to make letter templates
¾"-1½" Thick softwood for signs

Router bits needed:

◼ Any bit that will cut a groove, such as a straight bit, veining bit, core-box bit, or V-groove bit

Router accessories needed:

◼ Pantograph attachment

Making Signs with Incised Lettering

1. Cut out the lettering templates.

You can rout signs without wooden templates. However, if you want your lettering to be precise—or if you're going to make a lot of signs—you'll need letter templates. Enlarge the alphabet you want to use to *twice* the size of the letters you wish to rout. (Most pantograph attachments reduce the template or pattern by 50%.) Trace the letters on ¼″ hardboard and cut out the *inside*. (See Figure 1.) Glue this template to a ¼″ blank. All blanks must be exactly the same size, and the letters must be positioned exactly the same on the blanks.

2. Rout the first letter.

Set up your pantograph and router according to the manufacturer's instructions. Mount a ½″ core-box bit in the router, and put the sign stock and the template for the first letter in place. Rout the letter, keeping the stylus of the pantograph pressed against the edges of the template.

3. Repeat for the remaining letters.

Put the template for the next letter in place, and move the sign stock the space of one letter. Rout the next letter as you did the first. Continue until you have completed the sign. (See Figure 2.)

Figure 1.   To make templates for incised lettering, cut out the inside of the letters.

Figure 2.   Rout the letters in the stock, tracing the inside edges of the template with the pantograph's stylus.

SAMPLE ALPHABET

Making Signs with Raised Lettering

1. Cut out the lettering templates.

Make templates for the letters as you did before, but this time cut out the *shapes* of the letters and glue them to the blanks. (See Figure 3.)

TIP If you're good with a jigsaw, you can cut the templates for both incised *and* raised letters at the same time. The scrap from cutting the raised letter template becomes the template for the incised letter.

Figure 3.   To make templates for raised lettering, cut the shape *of the letters.*

2. Rout the stock around the first letter.

Set up the pantograph, sign stock, and template as you did before. Mount a ⅜″ straight bit in the router. Outline the first letter, tracing the edges of the template with the pantograph's stylus. Then work outward, routing away more and more stock. Be careful not to rout away too much— you don't want to remove stock that you'll need for the subsequent letters.

3. Repeat for the remaining letters.

Put the template for the next letter in place, and move the sign stock the space of one letter. Rout the stock around the next letter as you did the first. Continue until you have completed the sign. When you're finished, remove all the templates and go back and rout away more material, if you wish. (See Figure 4.)

Figure 4.   Outline the letter first, tracing the shape of the letter with the stylus. Then work outward from the letter, removing more and more stock with the router.

Making the Letters Stand Out

There are several things you can do to your signs to help the letters stand out from the background. Here are some suggestions:

A. Laminate two pieces of contrasting woods. Before you rout the signs, laminate two pieces of contrasting woods, such as maple and walnut. The top piece should be no more than ⅛″-¼″ thick. When you rout the sign, the contrasting wood underneath the top piece will be revealed.

B. Paint the wood. Before you rout the wood, paint the top surface. This paint should be either darker or lighter than the wood beneath it. When you rout the wood, the router will remove the thin layer of paint, revealing the wood underneath.

You can also paint the wood *after* you rout it. Let the paint dry, then sand the sign with a belt sander. The sander will remove the paint from the high areas, and leave the paint in the low spots.

C. Texture the background. Contrasting textures also set the background apart from the letters. There are two ways to texture the letters: (1) With a gouge, dig out the background area to give the background a hand-carved look. (See Figure 5.) Or, (2) punch little depressions all over the background with a hammer and a large nail. This gives the background a 'stippled' look. (See Figure 6.)

Figure 5. *Texture the background with a gouge, to give it a hand-carved look.*

Figure 6. *Punch the background with a hammer and nails to give it a 'stippled' look.*

TIP To help the stippling go faster, cut the heads off several nails and tape them together.

Jack Rabbit

Wooden animal silhouettes have always been a popular design motif. Gift shops carry the shapes of cats, ducks, geese, and sheep, just to name a few. Here's one more for the knickknack shelves: a jack rabbit. This particular wooden sculpture is built up to give the rabbit some contours, and the head pivots so you can arrange him in a variety of different poses.

This project shows off the capabilities of the router as a production tool. If you're just going to make one of these rabbits, you would probably use a bandsaw or a jigsaw. But when you're making two or more, you can use the router and the pin arm on the Router Table to duplicate the parts quickly and easily.

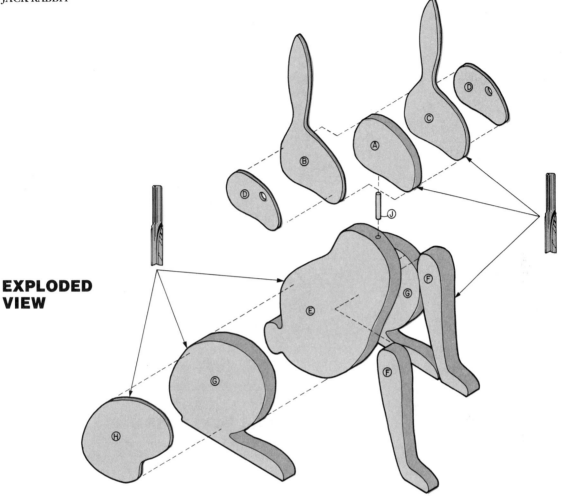

EXPLODED VIEW

Tools and Materials

Wooden parts*:

A.	Head	½″ x 3½″ x 5½″		
B.	Right ear	¼″ x 5″ x 9″		
C.	Left ear	¼″ x 5″ x 9″		
D.	Cheeks (2)	¼″ x 3″ x 4″		
E.	Body	1″ x 9″ x 10½″		
F.	Front legs (2)	1″ x 4″ x 7½″		

G.	Back legs (2)	1″ x 6½″ x 8½″
H.	Hips (2)	¼″ x 6″ x 6½″
J.	Dowel	⅜″ dia. x 1½″

*Measurements given are the size of
the blanks needed to cut the parts.

Router bit needed:

- ¼″ Straight bit

Making the Jack Rabbit

1. **Make a prototype.**   Enlarge the patterns for parts of the rabbit, trace them on scrap wood, and cut them out on a bandsaw. Sand the edges to remove any saw marks. Temporarily assemble the prototype with double-faced tape ('carpet' tape) to make sure all the parts are properly proportioned. When you're satisfied with the prototype, disassemble it.

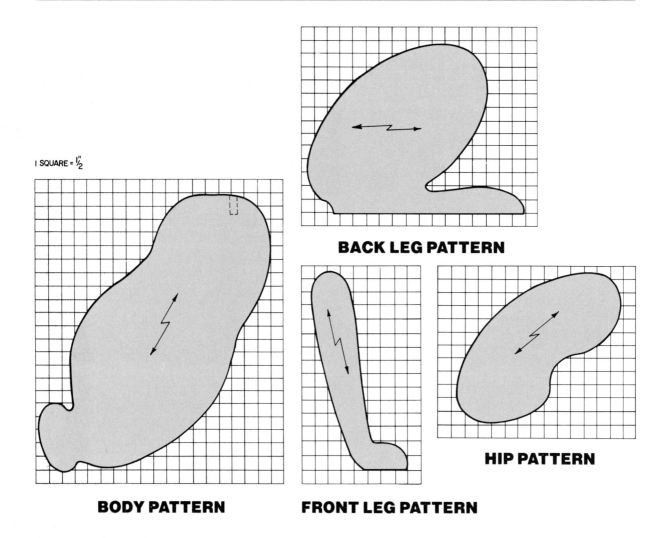

I SQUARE = 1½"

BACK LEG PATTERN

BODY PATTERN

FRONT LEG PATTERN

HIP PATTERN

Glue the parts of the prototype to rectangular blanks of ¼″ hardboard to make the templates. (See Figure 1.) Since some of the parts are the same, you need only make one template for those parts. For example, the two front legs are identical. Make just one template for both legs, using the best front leg from the prototype.

2. **Make the template with the prototype parts.**

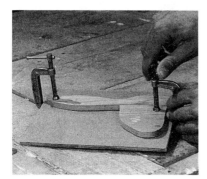

Figure 1. *Glue the prototype parts to ¼" thick hardboard blanks to make pin routing templates.*

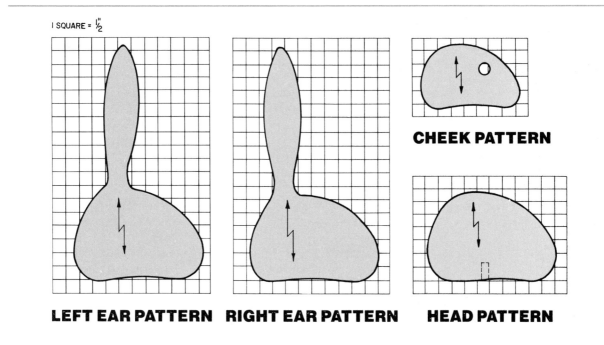

I SQUARE = ½″

LEFT EAR PATTERN   **RIGHT EAR PATTERN**   **HEAD PATTERN**

CHEEK PATTERN

3. **Pin rout the parts.**

Mount a ¼″ straight bit in the router, and mount the pin arm attachment on the Router Table. Adjust the position of the pin so that it's directly over the bit. There should be a space between the bit and the pin, so that the pin will contact the template, but not the stock you will attach to the template. (See Figure 2.)

Raise the bit so that it's ⅛″-¼″ above the table. Mount a wooden blank to one of the templates, using brads or double-faced tape. Carefully rout the shape of the part, keeping the template pressed against the pin. (See Figure 3.) When you've finished, raise the bit another ⅛″-¼″ and rout the shape a second time, again using the pin and the template to guide work. Repeat until you've cut completely through the blank. Be careful not to cut through the template.

> **TIP** You'll find it much easier to cut wooden shapes with the router if you make your cuts in several passes, routing slightly deeper with each pass. You can cut up to ¼″ deeper with each pass if you're working with softwood, but limit your cuts to ⅛″ if you use hardwood.

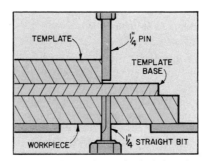

Figure 2. *Adjust the height of the pin so that it rests against the template while the bit is cutting the stock.*

Figure 3. *Rout the parts for the rabbit, using the pin to guide your work. Cut each part in several passes, cutting ⅛″-¼″ deeper with each pass.*

4. **Assemble the parts and finish the Jack Rabbit.**

Finish sand all parts. Glue the head, right ear, left ear, and cheeks together to make the head assembly. Glue the body, front legs, back legs, and hips together to make the body assembly. Glue a ⅜″ dowel in the hole at the top of the body assembly, but do *not* glue the other end in the head assembly. Instead, simply press the head in place on the dowel for a 'friction fit'. This will allow you to turn the head. Paint or finish the assembled project to suit your fancy, and tie a ribbon or bandanna around the neck.

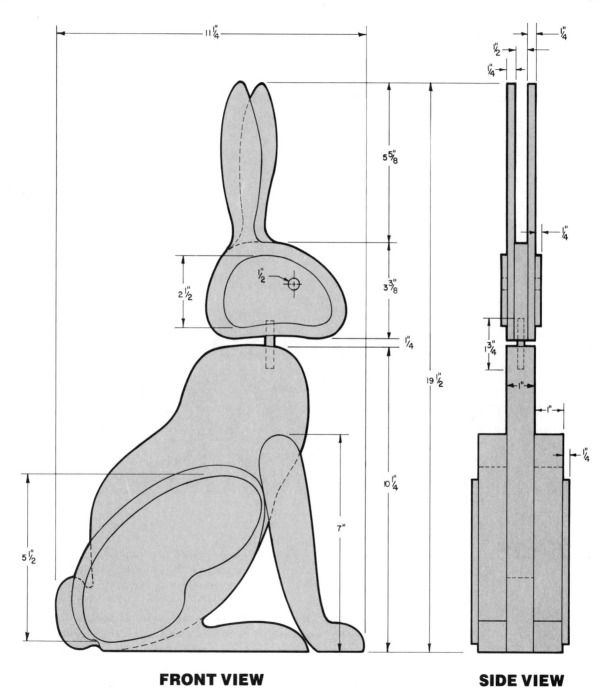

FRONT VIEW

SIDE VIEW

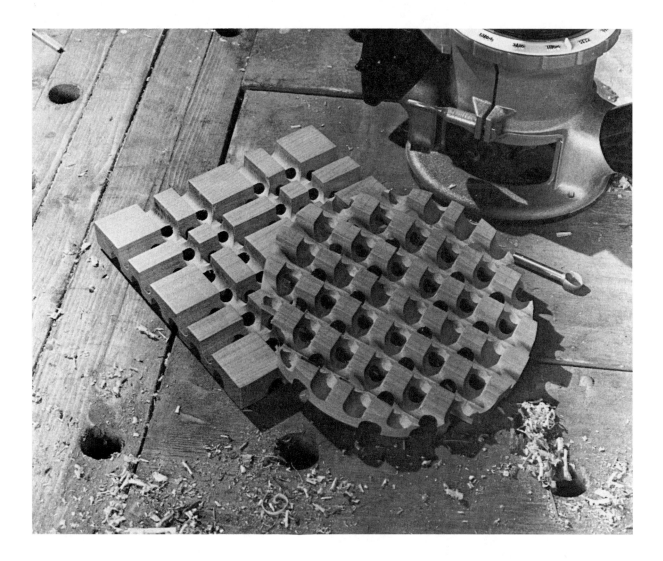

Trivets

Now and then you find a special woodworking technique that allows you to make a finished project in one simple operation. 'Piercing' with the router and the Router Table is just such a technique. The intricate trivets you see here were made from a single block of wood, using this unique method—and nothing else.

'Piercing' is cutting part way through a board with a blade or cutter, then turning the stock over and cutting again at a different angle. Where the two cuts intersect, the wood will be 'pierced'. Make multiple cuts on each side of the wood, and the pierced holes will form a pattern.

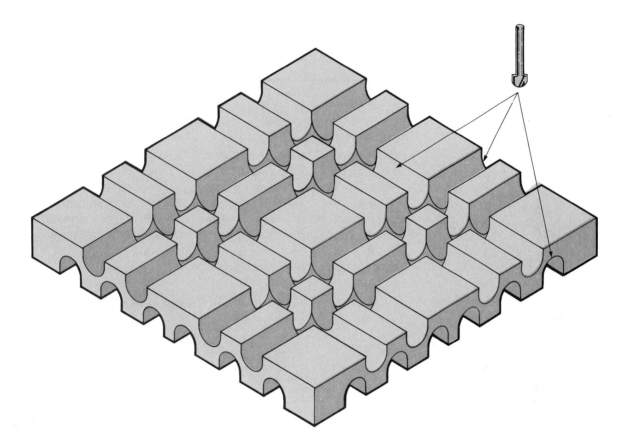

ISOMETRIC VIEW

Tools and Materials

Wooden parts:

A. Trivet   ¾″ x 6¾″ x 6¾″

Router bit needed:

- ½″ Core-box bit

Piercing the Trivets

1. Cut the blanks to size.

You'll want to make several trivets at a time—perhaps a dozen or more, for gifts. Cut as many blanks as you feel you need.

2. Cut the grooves in the blanks.

Mount a ½″ core-box bit in the router, and adjust the height so that the bit's ⁷⁄₁₆″ above the table—more than half the thickness of the trivet stock. Set the fence a short distance away from the bit. The dimensions in the working drawings are just examples; if you wish, you can use your own. With the fence guiding the stock, cut a groove. (See Figure 1.) Rotate the stock 90° and cut another. Repeat until you've cut four intersecting grooves in one face of the stock. Move the fence and cut another set of grooves. When you've cut all the grooves in one face, turn the trivet over and cut grooves in the other face. Where the grooves on one face intersect the grooves on the other, the wood will be 'pierced'. These pierced holes will form a pattern.

TIP As an option, you can also use straight, V-grooves, or veining bits to cut the grooves. Use the circle-cutting pin to make round trivets, after you've cut the grooves.

Figure 1. Using the fence as a guide, cut grooves in the trivet stock with a core-box bit.

3. Sand and finish the trivets.

Finish sand the sides and faces of the trivets, then apply an oil finish.

SIDE VIEW

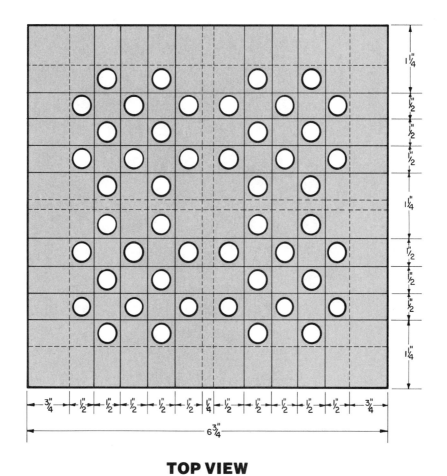

TOP VIEW

Magazine Basket

Like to do a little light reading from time to time? Here's a way to keep your magazines neatly arranged beside your easy chair. This magazine basket is small; it doesn't take up a great deal of floor space, but it's designed to hold over two dozen magazines.

This same basic design can be easily changed to store other things besides reading material. Add two more rails on the outside, so that there's less space between rails, and this makes a good knitting basket. Add extra rails on the outside, but omit the middle rails, and it becomes a place to store toys. Make other adjustments to store sewing notions, bottles and jars, records, photo albums, and a dozen other items.

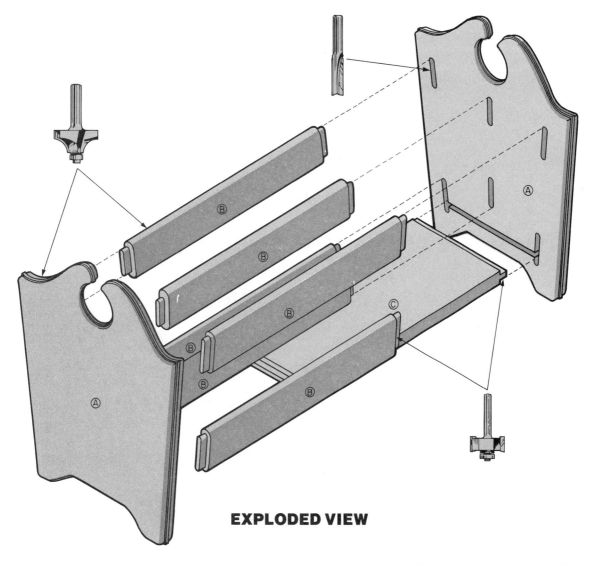

EXPLODED VIEW

Tools and Materials

Wooden parts:

A. Sides (2)   ¾″ x 10″ x 13″
B. Rails (6)   ¾″ x 2″ x 13½″
C. Bottom   ¾″ x 6⅜″ x 13½″

Router bits needed:

- ¼″ Straight bit
- Rabbeting bit
- ⅜″ Piloted quarter-round bit

Making the Magazine Basket

1. **Cut all parts to size.**	Cut all parts to size, and cut the straight *tapers* in the sides. Rip the edges of the bottom at 5°. Do not cut the shapes of the feet or the top just yet.
2. **Cut the joinery in the sides.**	Drill ¼″ holes, ⅜″ deep to mark the beginnings and the ends of the blind dadoes in the sides. With a ¼″ straight bit, rout these dadoes. (See Figure 1.) Clamp a straightedge to the side to guide the router.
3. **Cut the shape of the side.**	Once you've cut all the dadoes in the sides, trace the shape of the feet and the tops on the stock. Cut out the shapes on a bandsaw.

Figure 1. *Drill holes at the beginning and the end of each dado, so you'll know when to start and stop routing. Clamp a straightedge to the side to guide the router.*

Figure 2. *Cut the tenons in the ends of the rails and the bottom in two passes. Cut one side of a board, then turn it over and cut the other side.*

With a rabbeting bit, cut tenons in the ends of the rails and the bottom. Make each tenon in two passes—cut one side of the board, then turn the board over and cut the other side. (See Figure 2.) Use a miter gauge to guide the work, and the fence to gauge the length of the tenons. After you've cut the tenons, round over the top and bottom edge to fit the dadoes.

4. **Cut tenons in the rails and the bottom.**

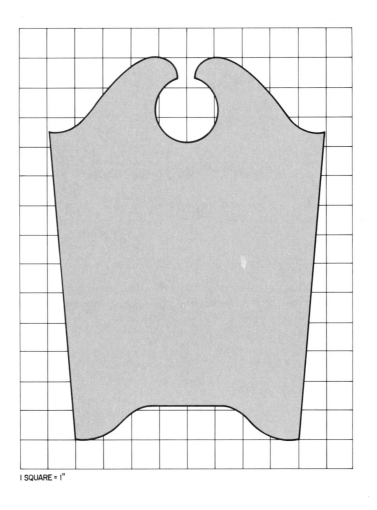

SIDE PATTERN

I SQUARE = I"

5. **Shape the rails and the sides.**

With a piloted ⅜″ quarter-round bit, round over all the edges of the sides and the rails, with the exception of the tenons. (See Figure 3.) If you wish, you can set the router to cut a ⅛″ **wide**, ¹/₁₆″ long 'step' in the rounded edge, as shown in the working drawings.

Figure 3. *Round over the edges of the rails and the sides with a piloted ⅜″ quarter-round bit.*

6. **Assemble and finish the basket.**

Finish sand all parts. Then assemble the sides, bottom, and rails with glue. After the glue dries, do any necessary touchup sanding, and apply a finish.

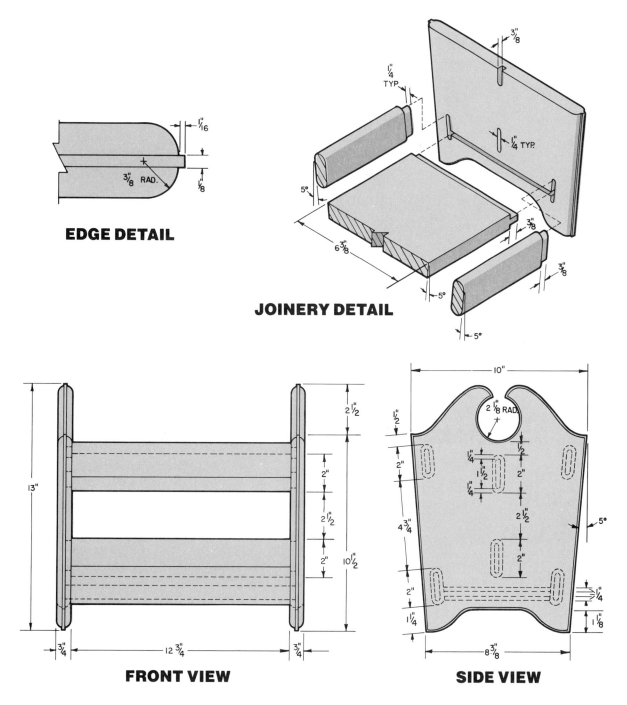

EDGE DETAIL

JOINERY DETAIL

FRONT VIEW

SIDE VIEW

Tissue Holder

A wooden tissue holder serves two purposes. The first is obvious: It makes the tissue box look elegant, adds a touch of class to something that is otherwise mundane. The second is less obvious, and more practical: It weights down the tissue box, helps it stay put. You're less likely to accidentally yank a heavy wooden tissue holder off a shelf when trying to extract a tissue.

The tissue box shown here is made with dovetails—the same joint you'd normally use to assemble a drawer. In this case, however, the joint is visible and adds a bit of decoration to the box. If you wish, you can enhance the effect by making the sides and the ends out of two *contrasting* species of woods (such as maple and walnut), so the dovetails are clearly visible.

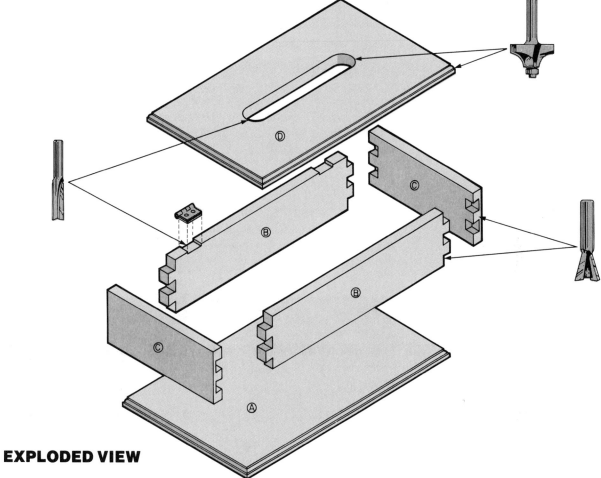

EXPLODED VIEW

Tools and Materials

Wooden parts:

A. Base	½" x 7" x 11½"	
B. Sides (2)	½" x 2½" x 10¼"	
C. Ends (2)	½" x 2½" x 6"	
D. Top	½" x 6½" x 11"	

Hardware:

- 1" x 1" Brass butt hinges and mounting screws

Router bits needed:

- ½" Dovetail bit
- ⅜" and ¾" Straight bits
- ¼" Piloted quarter-round bit

Router accessories needed:

- ⁷⁄₁₆" and ⅝" Guide collars
- Dovetail template

Making the Tissue Holder

1. **Cut the dovetails that join the sides and ends.**

Cut all stock to size. With a ½″ bit, a ⁷⁄₁₆″ guide collar, and a dovetail template, cut the half-blind dovetails that join the sides and the ends. (See Figure 1.) Since the stock is just ½″ thick and most dovetail templates are made for ¾″ stock, you will probably have to shim the parts with scraps of ¼″ wood. A few templates are engineered to compensate for different thicknesses of wood. Check the instruction manual that came with the template.

2. **Mortise the sides for hinges.**

Mount a ¾″ straight bit in the router. Carefully measure the thickness of your *closed* hinge, and adjust the height of the bit above the worktable to this measurement.Using a miter gauge to help hold and guide the stock, cut the mortises for the hinges. (See Figure 2.)

Figure 1. Cut the dovetail in the sides, using a dovetail template. Since most templates are made for ¾″ stock, you may have to adjust the template or shim the stock.

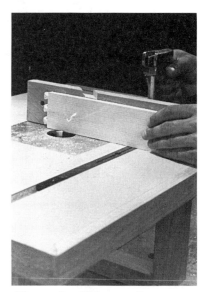

Figure 2. Using a miter gauge to hold and guide the stock, cut mortises in the side for hinges.

Mount a ⅝″ guide collar and a ⅜″ straight bit in the router. With a saber saw, make a template to cut the slot from ½″ plywood. The slot in the template should be ¼″ longer and ¼″ wider than the dimensions for the slot you want to cut in the top. Mount the template and drill a starting hole in the top. Place the router in position, with the bit in the hole, and turn it on. Rout the slot, keeping the guide pressed against the template. (See Figure 3.)

3. **Cut the slot in the top.**

Figure 3. *Cut the slot in the top with a straight bit, using a template to guide the cut. Make the cut in several passes, cutting ⅛″-¼″ deeper with each pass.*

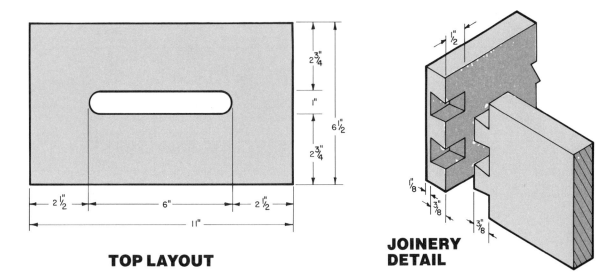

TOP LAYOUT

JOINERY DETAIL

4. Shape the top and the base.

With a ¼″ quarter-round bit, round over the edges of the slot in the top. (See Figure 4.) Use this same bit to shape the edges of the top and the base.

Figure 4. Round the edge of the slot in the top with a quarter-round bit.

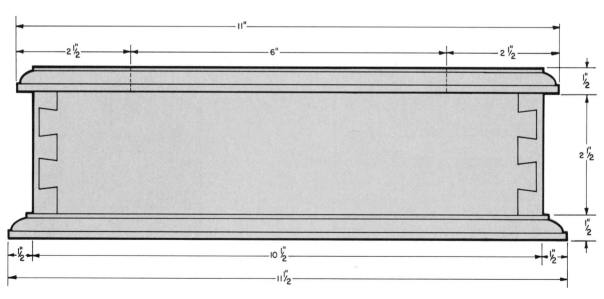

FRONT VIEW

Assemble the sides and ends with glue. After the glue cures, finish sand all parts. Attach the base to the box with screws, and hinge the top to the side that you mortised. Finally, apply a finish.

5. Assemble and finish the box.

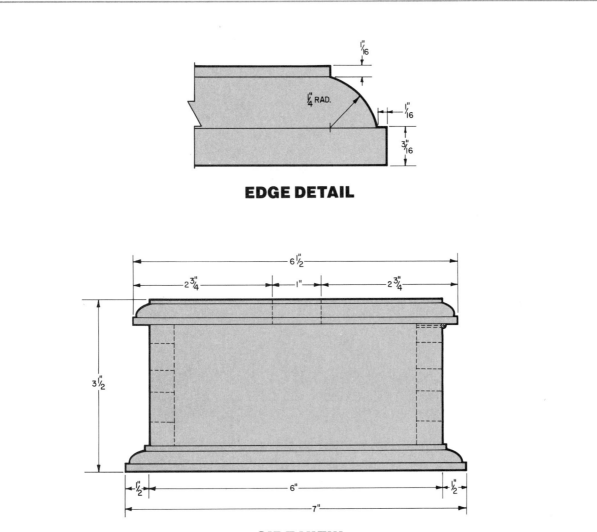

EDGE DETAIL

SIDE VIEW

Gumball Machine

Gumball machines hold a fascination for children and grown-ups alike, not just because they dispense candy, but also because of the way they work. Gumball machines are great 'contraptions', and as such, it's a challenge to try to figure out how they dispense the balls. Give a kid a gumball machine and enough pennies, and he'll probably empty the machine to watch it work, long before he eats all the gumballs.

Here's a gumball machine whose source of fascination is its simplicity! There is only one moving part—the jar that holds the gumballs. That makes this project easy to build. But it's still just as much fun to watch it work.

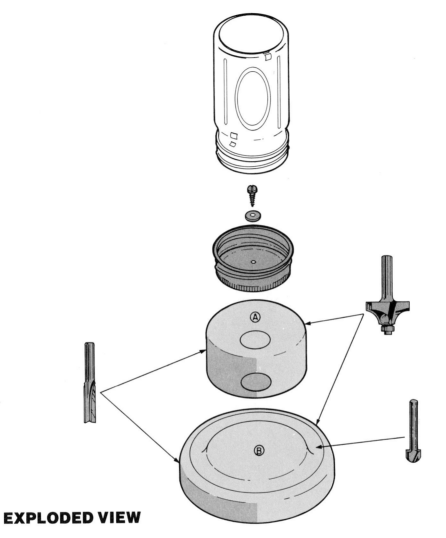

EXPLODED VIEW

Tools and Materials

Wooden parts:

A. Base   6¼″ dia. x ¾″
B. Body   4″ dia. x 1½″

Hardware:

- One-pint glass canning jar with a screw-on lid

Router bits needed:

- ½″ Straight bit
- ½″ Core-box bit
- ⅜″ Quarter-round piloted bit

Making the 'Machine'

1. **Cut the base and the body.**

Cut the base so that it's perfectly round. You can do this on a bandsaw or a jigsaw, but you can also do it on your Router Table. Mount a ½″ straight bit in the router and adjust the circle-cutting pin on the table so that it's 3⅛″ from the *edge* of the bit. Mount the stock so that it pivots on the pin, and cut the circle. (See Figure 1.) Take small bites, cutting just ⅛″ deeper with each revolution.

Move the pin so that it's 2″ away from the *edge* of the bit, and cut the body. You'll have to cut the body in two parts, each ¾″ thick. Most straight bits aren't long enough to cut a piece of wood 1½″ thick. When you've cut both parts of the body, glue them together. (See Figure 2.)

You'll find there is a unique advantage to using a router to cut circles. Unlike bandsaws and jigsaws, the router leaves no millmarks. The edge will be perfectly smooth and round.

Figure 1. *Cut the base on the Router Table, using a straight bit and the circle-cutting pin.*

Figure 2. *To make the body, cut two ¾″ thick circular pieces and glue them together.*

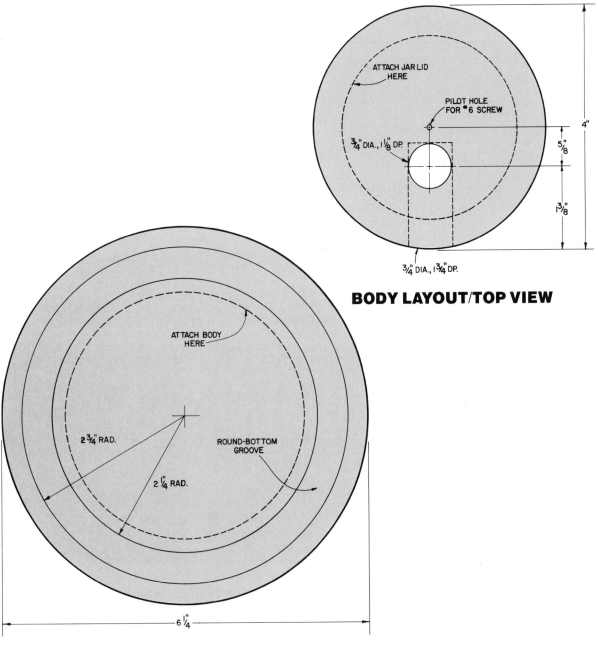

BODY LAYOUT/TOP VIEW

BASE LAYOUT/TOP VIEW

2. Cut the groove in the base.

Mount a ½″ core-box bit in your router, and adjust the circle-cutting pin on the Router Table so that it's 2½″ from the *center* of the bit. Mount the base so that it pivots on the pin, and cut a round-bottom groove in the base, ½″ wide and ¼″ deep, all around the circumference. (See Figure 3.)

3. Drill the body to create the 'delivery chute' for the gumballs.

Using a square, mark the locations of the entrance and exit for the gumballs on the body. Drill the exit first—the hole in the side of the body. To do this, clamp the body to a scrap of 2 x 4. This will keep the body from rocking as you drill the hole. Then tilt the table of your drill press 10° and drill the hole. (See Figure 4.) The hole must slant *up* from the exit, as shown in the working drawings.

Remove the 2 x 4 from the body and drill the entrance hole in the top surface. This hole should meet with the exit hole, creating a 'delivery chute' for the gumballs.

Finally, drill a small pilot hole in the top surface, smack in the center. You'll use this pilot hole later on to mount the jar lid to the base assembly.

Figure 3. *Cut the groove in the base on the Router Table, using a core-box bit and the circle-cutting pin.*

Figure 4. *Drill the exit hole in the body at a 10° angle, so that the delivery chute will slant up from the exit. Clamp a 2 x 4 to the body to keep it from rocking while you're drilling the hole.*

Set up your router table as a shaper. Mount a piloted ⅜″ quarter-round bit in the router, and round over the top edges of the base and the body. (See Figure 5.)

Figure 5.   Round over the top edge of the base and the body with a quarter-round bit.

4. **Round over the top edge of the base and the body.**

Finish sand the base and the body, then glue the body to the base. Make sure that the body is perfectly centered on the base. Remove any glue squeeze-out with a wet rag.

5. **Glue the body to the base.**

After the glue sets up, finish sand the assembly, if it needs it. Then apply a finish. We suggest using a non-toxic finish, since this project will be used to dispense candy. You have several choices: If you want a quick finish, rub the assembly down with mineral oil or 'salad bowl dressing'. This dressing is available through most woodworking supply companies. If you're in no hurry, use Danish oil. Danish oil has a low toxicity after 72 hours, and becomes completely non-toxic after several weeks.

6. **Finish the base assembly.**

7. Drill the jar lid.

Mount a metal-cutting drill bit in your drill press, and drill the delivery hole in the jar lid, where shown in the drawings. Also, drill the pivot hole in the center of the lid.

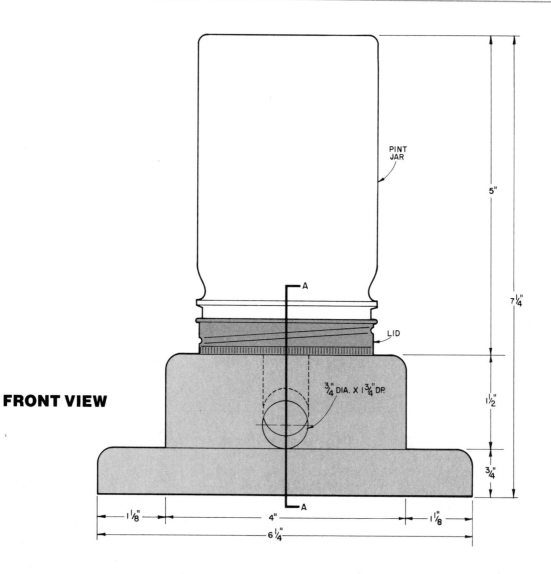

FRONT VIEW

PINT JAR

LID

$\frac{3}{4}$" DIA. X 1$\frac{3}{4}$" DP.

5"

7$\frac{1}{4}$"

1$\frac{1}{2}$"

$\frac{3}{4}$"

1$\frac{1}{8}$"

4"

1$\frac{1}{8}$"

6$\frac{1}{4}$"

Attach the jar lid to the base assembly with a roundhead wood screw and flat washer. Tighten the screw down so that it holds the lid snug against the body, but not tight. The jar lid must be able to pivot freely on the screw. Fill the jar with gumballs and screw it to the jar lid. To dispense a gumball, simply turn the jar so that the hole in the lid lines up with the entrance hole of the delivery chute. One or two gumballs will drop down. To prevent any more from dropping down, give the jar another half turn.

8. **Attach the jar lid and the jar to the base assembly.**

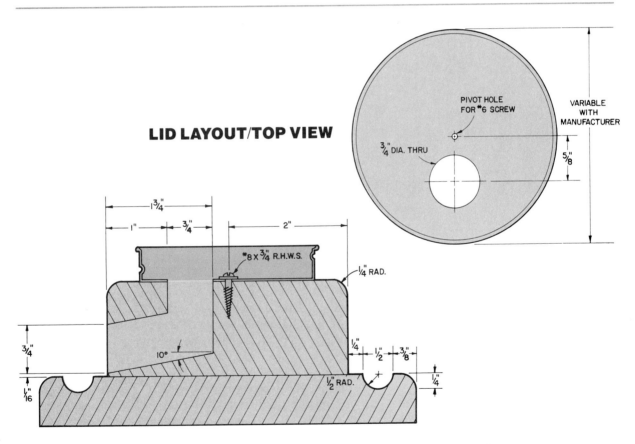

LID LAYOUT/TOP VIEW

PIVOT HOLE FOR #6 SCREW

VARIABLE WITH MANUFACTURER

$\frac{3}{4}$" DIA. THRU

$\frac{5}{8}$"

$1\frac{3}{4}$"

1" $\frac{3}{4}$" 2"

#8 X $\frac{3}{4}$" R.H.W.S.

$\frac{1}{4}$" RAD.

$\frac{3}{4}$"

$10°$

$\frac{1}{4}$" $\frac{1}{2}$" $\frac{3}{8}$"

$\frac{1}{16}$"

$\frac{1}{2}$" RAD.

$\frac{1}{4}$"

SECTION A

Kitchen Island

Need some more storage space in your kitchen? More counter space? Here's a possible solution to both problems: a rolling kitchen island. This movable cabinet has drawers and adjustable shelving. Two slots in the top will hold knives. And if you need even more counter space, you can add an optional drop leaf behind the cabinet.

Most of the joinery on this cabinet is done with a router. You use it to make dadoes and rabbets to assemble the case, tongues and grooves to assemble the frames, spline joints to attach the trim, dovetails to make the drawer, and cabinet lips to fit the doors. You even use the router to trim the laminate. In fact, the only power tools you need besides a router are a saw and a drill.

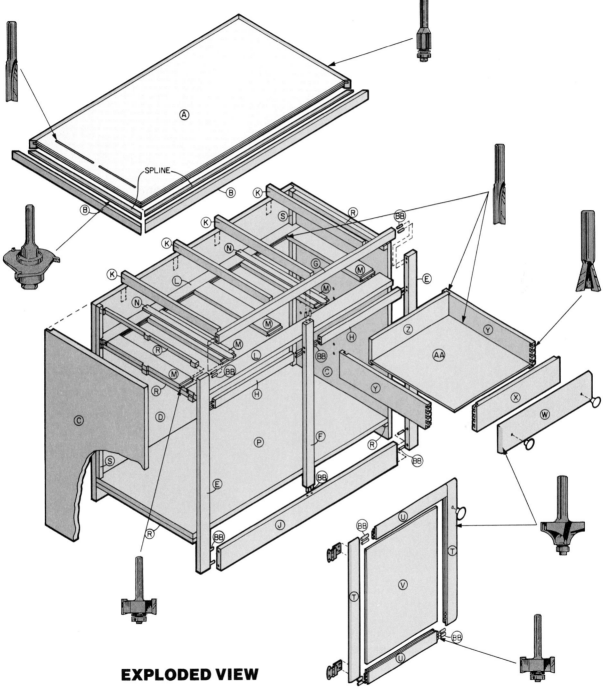

SPLINE

EXPLODED VIEW

Tools and Materials

Wooden parts:

A. Top   ¾" x 22½" x 43½"
B. Top trim
   (total)   ¾" x 1¼" x 144"
C. Sides (2)   ¾" x 21¼" x 34¾"
D. Back   ¾" x 34¾" x 42¼"
E. Outer front
   frame
   stiles (2)   ¾" x 2' x 34¾"
F. Middle front
   frame stile ¾" x 1½" x 29½"
G. Upper front
   frame rail   ¾" x 1½" x 38½"
H. Middle front
   frame
   rails (2)   ¾" x 1½" 18½"
J. Lower front
   frame rail   ¾" x 3¾" x 38½"
K. Kickers
   (4)   ¾" x 1½" x 20½"
L. Web frame
   rails (2)   ¾" x 2" x 41"
M. Web frame
   stiles (5)   ¾" x 3" x 17¼"
N. Drawer
   guides (2)   ¼" x 1" x 19¾"

P. Bottom
   shelf   ¾" x 20½" x 41"
Q. Adjustable
   shelf   ¾" x 14" x 40⅞"
R. Horizontal
   cleats (6)   ¾" x ¾" x 20½"
S. Vertical
   cleats (2)   ¾" x ¾" x 24¾"
T. Door frame
   stiles (4)   ¾" x 2" x 24¾"
U. Door frame
   rails (4)   ¾" x 2" x 16"
V. Door frame
   panels
   (2)   ¼" x 15⅞" x 21⅜"
W. Drawer
   faces (2)   ⅜" x 4¾" x 19¼"
X. Drawer fronts
   (2)   ¾" x 3¹⁵/₁₆" x 18⁷/₁₆"
Y. Drawer
   sides (4)   ¾" x 3¹⁵/₁₆" x 19⅝"
Z. Drawer backs
   (2)   ¾" x 3¹⁵/₁₆" x 17¹¹/₁₆"
AA. Drawer bottoms
   (2)   ¼" x 17⁹/₁₆" x 18⅝"
BB. Dowels (38)   ⅜" dia. x 2"

Hardware:

■ 1½" pulls (6)
■ Offset, self-closing cabinet hinges, with mounting screws (2 pair)
■ Shelving supports (4)
■ 4" Heavy-duty caster, with mounting screws (4)
■ #8 x 1¼" Flathead wood screws (60-72)
■ #8 x 1" Flathead wood screws (8-12)

Router bits needed:

■ ¼" Straight bit
■ Rabbeting bit
■ ½" Dovetail bit
■ ¼" Piloted quarter-round bit

Router accessories needed:

■ ⁷/₁₆" Guide collar
■ Dovetail template

Making the Kitchen Island

1. Cut all the parts.

Carefully check the dimensions in the list of wooden parts against the measurements on the working drawings, then cut all the parts to size.

2. Cut the joinery in the sides and web frame parts.

With a rabbeting bit, cut the rabbets in the sides. (See Figure 1.) You can also use this bit to make the tongues in the web frame stiles. (See Figure 2.) To cut the grooves in the rails, use a ¼" straight bit. (See Figure 3.) Assemble the web frame with glue.

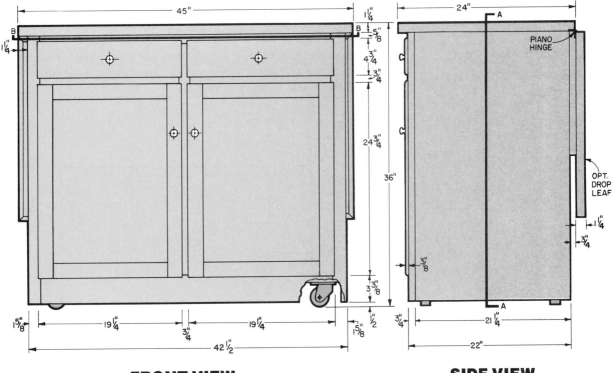

FRONT VIEW

SIDE VIEW

Figure 1. Cut the rabbets in the sides with a rabbeting bit.

Figure 2. You can also use the rabbeting bit to make the tongue in the frame parts.

Figure 3. To cut the grooves, use a ¼" straight bit. The fence guides the work.

3. Make the front frame.

Unlike the other frames in this project, the front frame parts are joined with dowels. Use a doweling jig to drill the dowel holes in the parts, then assemble the front frame with glue.

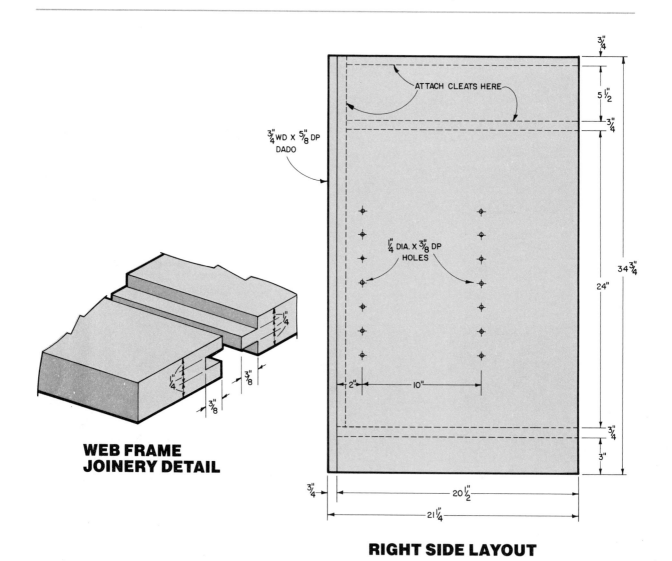

ATTACH CLEATS HERE

$\frac{3}{4}$" WD X $\frac{5}{8}$" DP DADO

$\frac{1}{4}$" DIA. X $\frac{3}{8}$" DP HOLES

$\frac{3}{4}$"

$5\frac{1}{2}$"

$\frac{3}{4}$"

$34\frac{3}{4}$"

24"

$\frac{3}{4}$"

3"

2"

10"

$\frac{3}{4}$"

$20\frac{1}{2}$"

$21\frac{1}{4}$"

$\frac{1}{4}$"

$\frac{1}{4}$"

$\frac{3}{8}$"

$\frac{3}{8}$"

WEB FRAME JOINERY DETAIL

RIGHT SIDE LAYOUT

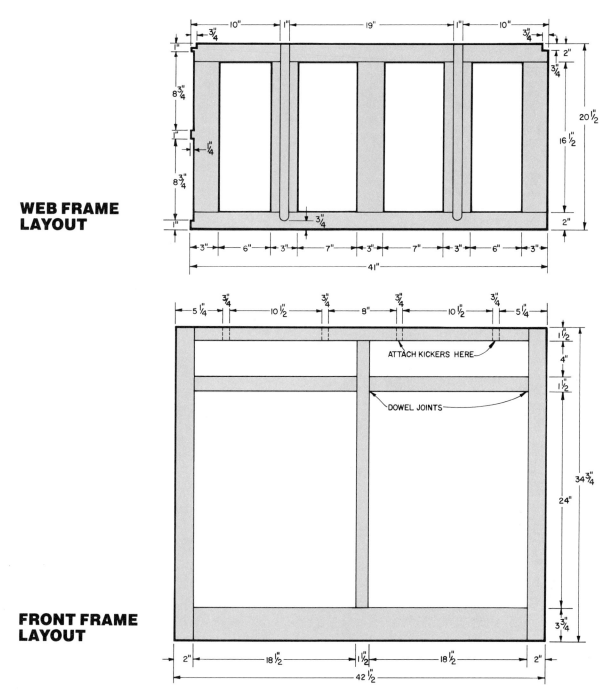

WEB FRAME LAYOUT

FRONT FRAME LAYOUT

ATTACH KICKERS HERE

DOWEL JOINTS

4. Assemble the case.

TIP To help move this project around your shop as you work on it, you may want to attach the casters to the case at this point.

While you're working with the drill, make the holes in the sides for the shelving support. Then attach the cleats to the sides with glue and screws. Assemble the sides, bottom shelf, back, and web frame as a unit—you'll probably need a helper for this operation. Then attach the front frame and kickers to the assembly. Countersink and counterbore the screws, then cover the heads with wooden plugs. Finally, attach the top.

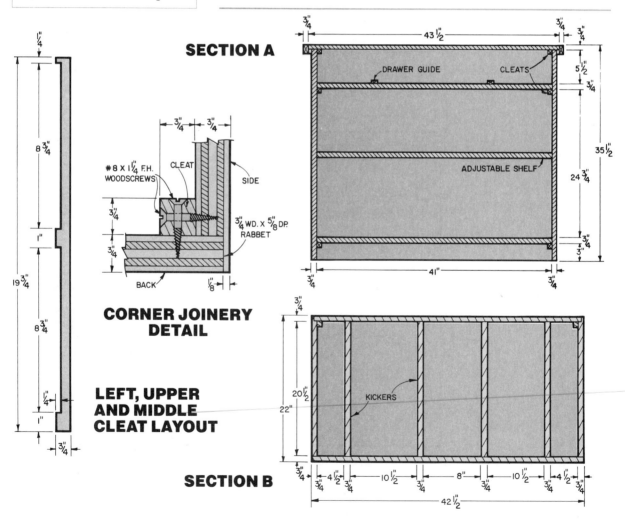

Cut the laminate slightly larger than the top, then bond it to the top with contact cement. Trim the laminate flush with the top with a straight trimmer bit. (See Figure 4.) *Option:* If you wish to add a drop leaf to your kitchen island, make the drop leaf at the same time you're making the top.

5. Laminate the top.

With a slotting cutter bit, cut ¼″ grooves for splines all the way around the top and in the inside edge of the trim. (See Figure 5.) Insert splines in the grooves, and attach the trim to the top with glue.

6. Attach the edge trim to the top.

If you wish, you can cut slots to hold knives in the laminated top. Use a straightedge to guide the router, and cut the slots with a ¼″ straight bit. (See Figure 6.)

7. Cut knife slots.

If you've elected to add the drop leaf to this project, make the supports at this time. Also, add ½″ thick ribs to the underside of the drop leaf, where it will rest on the supports.

8. Make the drop leaf supports (optional).

Figure 4. *Trim the laminate to its final dimensions with the router. Use a straight trimmer bit.*

Figure 5. *Use a slotting cutter bit to make the grooves for the splines. Cut spline grooves in both the top and the trim.*

Figure 6. *If you wish, cut slots in the top to store kitchen knives. Use a straightedge to guide the router, and a ¼″ straight bit to make the slot.*

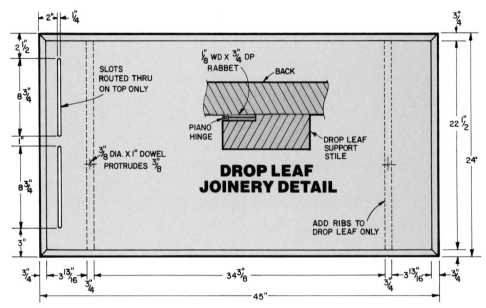

2" WD X 3/4" DP
RABBET

BACK

PIANO
HINGE

DROP LEAF
SUPPORT
STILE

**DROP LEAF
JOINERY DETAIL**

SLOTS
ROUTED THRU
ON TOP ONLY

3/8" DIA. X 1" DOWEL
PROTRUDES 3/8"

ADD RIBS TO
DROP LEAF ONLY

TOP AND DROP LEAF LAYOUT

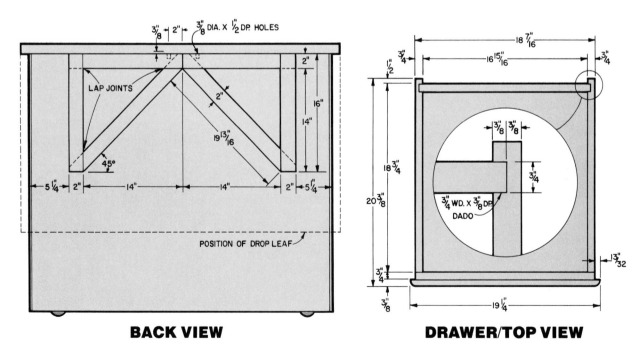

3/8" DIA. X 1/2" DP. HOLES

LAP JOINTS

45°

19 13/16"

POSITION OF DROP LEAF

BACK VIEW

3/4" WD. X 3/8" DP
DADO

DRAWER/TOP VIEW

Cut the joinery in the drawer parts. Use a ¾″ straight bit to make the dadoes in the drawer sides. Change to a ½″ dovetail bit, and use a guide collar and a dovetail template to make the half-blind dovetails that join the sides to the fronts. (See Figure 7.) Cut the grooves for the drawer bottom with a ¼″ straight bit. Finally, round over the edges of the drawer face with a ¼″ quarter-round bit. Notch the bottom edges of the backs, and assemble the drawers with glue. Reinforce the glue joints between the drawer faces and the drawer fronts with screws, driven from the *inside* of the drawers.

9. **Make the drawers.**

The door frames are made in a similar manner to the web frame—cut the tongues with a rabbeting bit and the grooves with a ¼″ straight bit. However, the joints are reinforced by dowels. Drill the dowel holes *before* you cut the tongues and grooves. (See Figure 8.) Assemble the door parts with glue; however, *do not* glue the panels in place. Let them float free in the grooves. When the glue has cured, cut a cabinet lip with the rabbeting bit and round over the lip with a ¼″ quarter-round bit.

10. **Make the doors.**

Figure 7. *Cut the dovetails that hold the drawer together with a dovetail bit, a guide collar, and a template.*

Figure 8. *The tongue and groove joinery on the door frames must be reinforced with dowels. Drill the dowel holes* before *you cut the tongues and grooves.*

11. Finish the cabinet.

Finish sand all parts, being careful not to scratch the laminate. Apply a finish to both the inside *and* the outside of the cabinet, the doors, and the drawers—this will help to keep the parts from warping. When the finish is dry, put the drawers in place. Hang the doors on the front frame with offset hinges. If you've added the drop leaf, attach the drop leaf and the drop leaf supports with piano hinges. Finally, attach pulls to the drawers and doors.

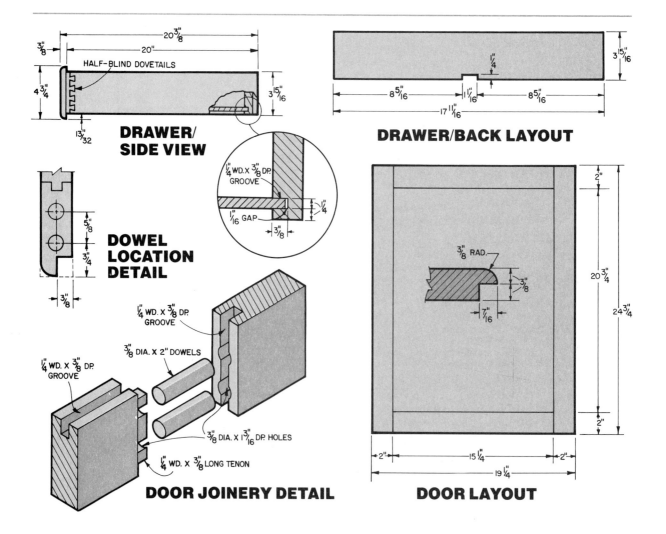

DRAWER/SIDE VIEW

DRAWER/BACK LAYOUT

DOWEL LOCATION DETAIL

DOOR JOINERY DETAIL

DOOR LAYOUT

METRIC EQUIVALENCY CHART

MM—MILLIMETRES   CM—CENTIMETRES

INCHES TO MILLIMETRES AND CENTIMETRES

INCHES	MM	CM	INCHES	CM	INCHES	CM
⅛	3	0.3	9	22.9	30	76.2
¼	6	0.6	10	25.4	31	78.7
⅜	10	1.0	11	27.9	32	81.3
½	13	1.3	12	30.5	33	83.8
⅝	16	1.6	13	33.0	34	86.4
¾	19	1.9	14	35.6	35	88.9
⅞	22	2.2	15	38.1	36	91.4
1	25	2.5	16	40.6	37	94.0
1¼	32	3.2	17	43.2	38	96.5
1½	38	3.8	18	45.7	39	99.1
1¾	44	4.4	19	48.3	40	101.6
2	51	5.1	20	50.8	41	104.1
2½	64	6.4	21	53.3	42	106.7
3	76	7.6	22	55.9	43	109.2
3½	89	8.9	23	58.4	44	111.8
4	102	10.2	24	61.0	45	114.3
4½	114	11.4	25	63.5	46	116.8
5	127	12.7	26	66.0	47	119.4
6	152	15.2	27	68.6	48	121.9
7	178	17.8	28	71.1	49	124.5
8	203	20.3	29	73.7	50	127.0

YARDS TO METRES

YARDS	METRES	YARDS	METRES	YARDS	METRES	YARDS	METRES	YARDS	METRES
⅛	0.11	2⅛	1.94	4⅛	3.77	6⅛	5.60	8⅛	7.43
¼	0.23	2¼	2.06	4¼	3.89	6¼	5.72	8¼	7.54
⅜	0.34	2⅜	2.17	4⅜	4.00	6⅜	5.83	8⅜	7.66
½	0.46	2½	2.29	4½	4.11	6½	5.94	8½	7.77
⅝	0.57	2⅝	2.40	4⅝	4.23	6⅝	6.06	8⅝	7.89
¾	0.69	2¾	2.51	4¾	4.34	6¾	6.17	8¾	8.00
⅞	0.80	2⅞	2.63	4⅞	4.46	6⅞	6.29	8⅞	8.12
1	0.91	3	2.74	5	4.57	7	6.40	9	8.23
1⅛	1.03	3⅛	2.86	5⅛	4.69	7⅛	6.52	9⅛	8.34
1¼	1.14	3¼	2.97	5¼	4.80	7¼	6.63	9¼	8.46
1⅜	1.26	3⅜	3.09	5⅜	4.91	7⅜	6.74	9⅜	8.57
1½	1.37	3½	3.20	5½	5.03	7½	6.86	9½	8.69
1⅝	1.49	3⅝	3.31	5⅝	5.14	7⅝	6.97	9⅝	8.80
1¾	1.60	3¾	3.43	5¾	5.26	7¾	7.09	9¾	8.92
1⅞	1.71	3⅞	3.54	5⅞	5.37	7⅞	7.20	9⅞	9.03
2	1.83	4	3.66	6	5.49	8	7.32	10	9.14